WIRELESS MARKETING

WIRELESS MARKETING

ROBERT A. STEUERNAGEL

A Wiley-Interscience Publication

JOHN WILEY & SONS, INC.

New York · Chichester · Weinheim · Brisbane · Singapore · Toronto

This book is printed on acid-free paper.

Copyright © 1999 John Wiley & Sons. All rights reserved.

Published simultaneously in Canada.

No part of this publication may be reproduced, stored in a retrieval system or transmitted in any form or by any means, electronic, mechanical, photocopying, recording, scanning or otherwise, except as permitted under Sections 107 or 108 of the 1976 United States Copyright Act, without either the prior written permission of the Publisher, or authorization through payment of the appropriate per-copy fee to the Copyright Clearance Center, 222 Rosewood Drive, Danvers, MA 01923, (978) 750-8400, fax (978) 750-4744. Requests to the Publisher for permission should be addressed to the Permissions Department, John Wiley Sons, Inc., 605 Third Avenue, New York, NY 10158-0012, (212) 850-6011, fax (212) 850–6008, E-Mail: PERMREQ@WILEY.COM.

This publication is designed to provide accurate and authoritative information in regard to the subject matter covered. It is sold with the understanding that the publisher is not engaged in rendering professional services. If professional advice or other expert assistance is required, the services of a competent professional person should be sought.

Library of Congress Cataloging-in-Publication Data:

Steuernagel, Robert.
    Wireless marketing / Robert Steuernagel.
        p. cm
    "Wiley-Interscience publication."
    Includes index.
    ISBN 0-471-31650-4 (alk. paper)
    1. Cellular telephone services industry.   2. Cellular telephone equipment industry.   3. Telephone, Wireless–Marketing.   4. Cellular telephones–Marketing.   5. Market surveys.   I. Title.
    HE9713.S74   2000
    384.5–dc21
                                                   99-16527
                                                    CIP

Printed in the United States of America.

10 9 8 7 6 5 4 3 2 1

CONTENTS

PREFACE

Any attempt to describe the wireless industry is like trying to describe a bullet as it passes by. While change is the watchword for the entire business environment, its effects are multiplied in an industry that is characterized by rapid growth and constant evolution in both technology and markets.

The wireless industry is one of the most exciting with which to be associated. It is one of the few industries that has all the elements required for maximum career satisfaction: first, an expanding industry with the potential for advancement, personal growth, and financial rewards; second, a high-technology area with all of the associated challenges, excitement, and promise; and third, a product that fills a genuine, urgent customer need.

The wireless industry is relatively naive in marketing, in spite of its phenomenal success. Perhaps its lack of marketing expertise is a legacy of its owners, mostly telephone companies, known for their lack of marketing prowess, or other communications and private companies whose marketing credentials are likewise not their best asset. The industry's lack of marketing ability has certainly not prevented it from gaining customers, but has caused it to pay an enormous price to get them. The conventional wisdom in the industry seems to harbor a sex-biased belief that "real men sell; others market," and that marketing is "fluff" that sales managers do in their spare time. Perhaps this book can fill a gap in the understanding of marketing techniques that can substantially lower the cost of acquiring and keeping wireless customers.

The purpose of this book is to demonstrate how intelligent marketing contributes to wireless success when combined with good sales techniques.

While it is written for marketing and sales managers and executives, it is important for managers at all levels and functions in the wireless industry to appreciate how marketing strategy influences, and is affected by, engineering, finance, operations, and so on. Its emphasis is on marketing strategies rather than sales strategies, recognizing that sales is just one important element of the marketing effort. It assumes that the reader has some knowledge of basic marketing techniques and shows how they can be applied effectively to wireless communications. A general understanding of wireless service and cellphones is also assumed.

Many of the statistics and examples used here will differ from the experience of readers. These statistics, such as usage per customer, are based on a mixture of my experience, industry averages, and actual results in specific markets and may also be outdated before they are printed. They should be treated as examples of benchmarks for marketing decisions, with the latest statistics for the reader's market(s) as the most appropriate operating data. My recommendations for wireless marketing operations are based on what has worked for me and what is currently working for others, combined with some things that might work for all in the future and an exclusion of things that have never worked.

No one has all the answers. If this book merely stimulates readers to study the marketing issues explored here more carefully and arrive at their own conclusions, I will have far exceeded my purpose.

ABOUT THE AUTHOR

Robert Steuernagel is an independent consultant and also an associate Vice President of The Strategis Group, wireless consultants in Washington, DC. He specializes in marketing and sales consulting for the wireless industry and has done international and domestic consulting for major wireless carriers.

He was most recently Vice President – Marketing and Sales for Authentix Network, Inc., a wireless fraud prevention company.

Mr. Steuernagel has served in executive positions for two of the United States' largest wireless carriers. He was Vice President–European Operations for Bell Atlantic International Wireless, in which he oversaw Bell Atlantic's ownership in wireless carriers in Italy and the Czech Republic, and served as Vice President of Marketing for Bell Atlantic Mobile, another Bell Atlantic division. He was Vice President of Marketing and Sales for AirTouch, and was also General Manager of Los Angeles Operations, one of the world's largest wireless markets, for that company.

Mr. Steuernagel is a veteran of the wireless industry, beginning with the original marketing and pricing strategies for cellular at AT&T's Advanced Mobile Phone Service (AMPS) prior to AT&T's divestiture in 1984. He has over 20 years' experience in telecommunications.

He is the author of *The Cellular Connection*, also published by John Wiley & Sons.

Mr. Steuernagel holds a BS from Bucknell University and an MBA from Fordham University. He lives in Southern California.

1

INTRODUCTION

THE WIRELESS MARKETPLACE

While wireless service came of age during a period of economic growth (from 1983 to 1988), its results (measured in customer growth) during the several recessionary and growth periods following have been as impressive as before.

The Cellular Telecommunications Industry Association (CTIA), the major industry association for wireless, reports that the number of wireless subscribers grew by almost 14 million in 1998, an annual rate exceeding 25%. This represents the highest annual gain ever recorded. It is truly extraordinary when you consider how much harder it is to maintain growth rates when the base is 55 million instead of 1 million a few years ago. While growth rates may not be as high as in previous years, the total growth in the market continues to climb.

Revenue was up over 20% for the year over the previous year. This indicates that while the revenue gain was healthy, the revenue per subscriber was lower for new customers than for the existing base.

The total domestic wireless subscriber base is now over 70 million, and the annual gross service revenue has grown to over 33 billion. The average monthly bill is $39.43. Merely dividing the total revenue by the total base (and again by 12) gives an estimate of monthly revenue per customer of $39.90.

The revenue per customer has been declining almost since wireless began, and figures going back to 1987 put the monthly figure at $96.83. It is not a phenomenon of economic conditions at all. While it is true that many newer markets have smaller areas and fewer customers with the highest need for wireless, large markets report that long-time customers use wireless as much as they always have; the primary reason for lower revenue per customer is that newer customers belong to market segments that find value in wireless, but do not need to use it as much as the earlier adopters. Declining wireless handset prices allow them to overcome entry barriers and subscribe to the service, even though they have a lower usage potential.

The outlook for wireless growth is just as healthy as its history (see Table 1.1). Originally projected to average 1% to $1\frac{1}{2}$% penetration in a metropolitan area, wireless penetration now exceeds 25%. Various estimates project wireless growing to over 100 million users (which would reflect about 40% penetration of the U.S. population) before the end of 2005. While much of this growth will include new entrants in personal communications services (PCS), cellular is expected to retain the largest portion of the market. This growth assumes continued deployment of new digital technologies, which will expand wireless' capacity, improve its performance, offer additional user benefits and services, and lower its costs.

The declining revenue per customer is just one statistic, like the price of wireless phones, that points to the rapid changes in the wireless marketplace over its short history. One of the challenges for wireless marketing, then, is to find ways to lower marketing costs and expand sales channels to newer market segments that produce lower revenue, while providing superior marketing and sales support to high-potential market segments. Such changes are already occurring in many markets.

TABLE 1.1   Wireless Growth

Year	Subscribers	Revenues (millions)
1984	91,600	178
1985	340,213	483
1986	681,865	853
1987	1,230,855	1,152
1988	2,069,441	1,959
1989	3,508,944	3,341
1990	5,283,055	4,549
1991	7,557,148	5,709
1992	11,032,753	7,823
1993	16,009,461	10,892
1994	24,134,421	14,230
1995	33,785,661	19,081
1996	44,042,992	23,635
1997	55,312,293	27,486
1998	69,209,321	33,133

Source: Cellular Telecommunications Industry Association.

WIRELESS MARKETING HISTORY

It is difficult to believe that there was a time when the market for cellular service was as uncertain as the market for advanced services like wireless data is today. As late as 1982, one year before introduction, it was not known whether carriers would sell at retail or wholesale, what price cellular customers would be willing to pay, and whether the target market for cellular would be defined by the dimensions of vertical markets (industries) or personal lifestyle demographics, such as income. Mobile telephones have been around for a long time, and many predicted that cellular would merely displace some of the 20,000 or so existing mobile telephones. These users were very unhappy with the service because of busy channels and poor quality.

However, cellular technology's increased dependability, ease of use, and capacity set the stage for the real reasons for cellular's success: aggressive new cellular carriers with only two permitted in each market, an enormous amount of free publicity, and the aggressive marketing of the service. All the marketing would have been for naught if the service had been poor. But the promise of cellular technology was fulfilled in commercial implementation, and early customers confirmed the value in cellular that was promised in the marketing.

Some of the marketing methods used today have a heritage in the business philosophy and regulatory constraints of the original participants. Before AT&T was broken up, the assumption was that it might be the only carrier and provide service only at wholesale to local mobile phone dealers. After it was decided that there would be two carriers, restrictions on the telephone companies' ability to sell customer premises equipment (CPE) or telephones in combination with services led the telephone companies to believe that retail sales agents would be required to provide both equipment and service subscription at a single sales point ("one-stop-shopping"). The agent concept has survived and flourished even after such restrictions were removed for a number of reasons we shall investigate.

Another regulatory constraint that helped shape cellular marketing was the reseller concept. Because the wireline[1] (WL) licensees of cellular were able to agree on market ownership early, while the diverse ownership of nonwire-line (NWL) carriers had much to settle before building systems, the Federal Communications Commission (FCC) agreed to require the WL carriers to resell cellular service at wholesale. This allowed the NWL company in each

---

[1]The FCC awarded two cellular licenses in each metropolitan or other geographic area: one to a telephone company affiliate ("wireline") and one to anybody else ("nonwireline"); the largest 90 markets were awarded on the merits of the application; the telephone companies agreed on which company would apply for each market rather than compete for them. The non-wirelines, as local, diverse business interests, could come to no such easy agreements.

market to sell service at retail by reselling the WL carrier's service until the NWL could complete its system and begin commercial service. This was known as the "headstart" provision. Its importance today is that it created an additional sales channel for wireless that all carriers might have foreclosed forever if left to their own judgment. They still often try to suppress it because of its historical relationship to supporting future carrier competitors, among other reasons.

Because cellular licenses were awarded on a market-by-market basis, there were no carriers or other providers with a large enough regional or national market to consider national advertising, sales or marketing strategy for wireless, except those that have been developed through consolidation or industry affiliation.

As a result of this environment, wireless has been until recently marketed and sold on a local basis; even large carriers such as AT&T Wireless and AirTouch decentralize their marketing as well as their management. While a decentralized approach is appropriate in today's business environment, it inhibits the development of broad strategies, devalues marketing skills and resources as a small appendage of the sales effort, and limits the ability to propagate successful local marketing programs from one market to another— and to communicate the learnings of failed ones. As a result, customer gain has been driven by the brute force of sales resources and special offers more than by the thoughtful application of broader marketing disciplines.

The legacy of this heritage is that the marketing of wireless is relatively unsophisticated. While wireless has achieved spectacular growth and acceptance, it has done so at an enormous marketing and sales cost (see Table 1.2). Wireless technology's lack of marketing sophistication is derived from its originally prescribed local orientation, the low marketing skills of its telephone company and other owners, and the short-term focus of decentralized management.

THE INDUSTRY

The wireless industry has at its core the two to eight carriers in each market and the independent, external sales channels that complement (or compete

TABLE 1.2   Estimated Marketing Cost per Wireless Activation

	Agent	Carrier
Sales cost	$250	$350
Loaded marketing cost	$400	$550

with) the carrier's own sales efforts. These external channels include resellers, agents, dealers, and retailers. The original complement of two cellular carriers in each city has been supplemented by two to six additional carriers with PCS licenses at higher frequencies. Also, carriers such as NEXTEL provide cellularlike service in different frequency bands.

The manufacturers of wireless handsets and network equipment are the second major element of the industry and influence the features of the service and the sales and value presentation to the customer.

Other major elements of the industry also influence its marketing: billing services providers and other vendors of products and services for carriers; makers of accessories to improve wireless communication and allow new applications of wireless service like FAX and data; and auxiliary service providers like voice messaging, paging, and local and long-distance telecommunications companies.

THE MARKETING FRAMEWORK

We will approach wireless marketing in a somewhat classical manner: to describe the product features and target markets and to show how the various elements of the marketing mix [the four P's: product, price, promotion, place (Distribution or sales)] interact with each other to maximize the penetration and satisfaction of the target market at appropriate cost levels. We will explore the fit of marketing with the overall strategic plan and look at the development of the marketing plan as part of the annual operations plan. We will look at marketing and sales organization and execution. Finally, we will discuss some particular problems of the marketing approach to wireless, some current trends in marketing strategies, and some guesses about the future as wireless is integrated further into the broader telecommunications environment.

RESTRICTIONS ON BELL COMPANIES

As mentioned, the wireline carriers (not just Bell companies) were required to offer service at wholesale so that the second carrier could market at retail until it had its own system. Wireline companies are still trying to get this requirement removed now that all systems are operating, indicating their desire to discourage resale.

The Bell companies were required to offer cellular or wireline CPE through a separate subsidiary from the carrier. This was eliminated around 1986.

The Bell companies were prohibited from providing information services until 1991.

The Bell companies were previously prohibited from providing long-distance service. The wireless divisions are now permitted to offer long distance.

The Bell companies were previously prohibited from manufacturing telephone equipment. Thus they could not manufacture, or have manufactured, wireless handsets. They were permitted to private label other manufacturers' designs.

2

MARKET AND PRODUCT DEFINITION

CUSTOMER BENEFITS

The original wireless customer saw benefits of wireless service in two major business-related categories: first, the ability to stay in touch for important decisions when away from traditional, landline telephones; and second, the ability to be more productive by converting time on the road (or other nonproductive time) into productive telephone time.

The evolution of wireless phones from primarily car-installed mobile phones to primarily portable handsets has occurred in a few short years; in parallel customers who originally described benefits in terms of automobile travel time can now with portable handsets envision a much larger set of places and circumstances in which wireless service can provide benefits. Today wireless can begin to compete with landline service as telephone users become more comfortable with having a portable phone as their main access line.

However, today's target market is no longer the business user. While the business user is the most attractive user, the preponderance of business users are already using wireless technology. An important marketing lesson we will observe in the wireless industry is the difference between the current customer base and the current target market. Too often we look at the current customer base for target-market demographic trends rather than who is currently buying.

If we look at classical marketing's consumer adoption process, in which potential customers go through stages of awareness, interest, evaluation, trial, and adoption, we can examine how the evaluation of benefits and other elements of the consumer's decision process relate to the marketing communications process. Table 2.1 shows some of the experiences of wireless customers as they go through the stages of the decision-making process. It is important that potential customers have these experiences to lead them to a purchase decision. The marketing communications that help this happen include advertising, which both educates the consumer and provides a time-limited special offer to act, and public relations materials such as press releases and community programs. As wireless technology has matured, the effectiveness of educational advertising has decreased as awareness of wireless through other means (e.g., use by associates) has increased. It is more important for the potential customer to have multiple exposures to different kinds of advertising in a single day or week, which can only occur when multiple channels and vendors promoting various offers of equipment, benefits, and service are active in the market.

Note that the adoption stage in Table 2.1 occurs *after* the customer has subscribed to the service. Subscriber-based services such as wireless must be

TABLE 2.1   The Stages of the Wireless Buying Process and the Related Experiences of the Potential Wireless Subscriber

Stage	Experiences of Potential Subscriber
Awareness	See cellphones used in TV and films
	See advertisements in all media
	Observe users on the street
	See wireless antennas on cars
Interest	Listen to wireless advertising and notice phones among other items advertised
	Notice that many peers seem to be using wireless when contacting them
	Have multiple exposures to users and ads in a single day
	Ask questions and actively seek information about wireless
Evaluation	Interpret benefits with reference to their own job and lifestyle
	See associates and friends with similar lifestyles adopt wireless services
	Experience specific occasions when they wish they had wireless service
	Inquire about a sales presentation
	Become delighted with certain models of wireless phones
Trial	Try it themselves through friends or sales presentations
	Buy under promotional offer
	Experience uncertainty over commitment to wireless
	Get "sticker shock" on receipt of first bill if price sensitive
Adoption	Express and see value to keep service
	Consider cost to keep service Opportunity cost to stop via (stranded investment in phone or service contract)
	Cannot function without it
	Wonder how they got along without it Change jobs or move

distinguished from products, in that products are sold only once, and the product is paid for up front (even if financed); with subscriber services, the customer pays only a portion of the lifetime cost for the service at a time and is given a chance every month to reevaluate whether the cost–benefit relationship justifies continued use (see the section titled "Services versus Products").

Therefore, as with other subscriber services and consumable repeat-purchase products, adoption actually occurs after a few months or after use allows the customer to evaluate the cost–benefit relationship.

MARKET SEGMENTATION

Early market research in wireless services uncovered a target market that was characterized as the mobile portion of business managers in certain vertical markets (industry categories). Only when cellular was actually offered commercially were the true driving factors characterizing the target market identified as a mix of personal lifestyle and business characteristics. Apparently it was not important that the potential customer worked in a certain geographic area of the country in a certain industry, although these characteristics correlated positively with wireless users.

It was found that the actual drivers of the original prime target market were not the type of industry nor the type of work, but the size of the company and the manager's relative position in the firm, combined with certain other demographic information. Thus, the profile of the highest potential customer for cellular was and perhaps still is the following:

Male
Age 28 to 55
Income over $70,000
Works in a company of less than 20 employees
Senior executive of the company
Owner of a luxury car

Some of these characteristics may be obvious, but the difficulty is to identify the characteristics that *cause* someone to be a wireless user rather than those that are merely correlated with users. The least obvious char-acteristics are the association of senior executives of very small companies, which apparently arises because these individuals need to take an active role in company field operations and stay constantly in touch for important decisions.

Market Changes

While these characteristics still hold for a small segment of the highest potential customer prospects for wireless services, few of these prospects remain (they are already customers). Moreover, target markets cannot be so narrowly defined. The entire target market has broadened, increasing the potential market, but making it more difficult to target individual segments. Half of wireless users are now women, and the best wireless customers are middle management of very large firms—a segment that was virtually nonexistent when wireless service was in its infancy. The range of age and income has broadened significantly, and the association with a luxury car is no longer a necessary prerequisite for a good wireless customer.

The declining price of wireless handsets has not only attracted a much broader base of business users but has put wireless services within the reach of consumers as well. The problem with the consumer market is that the cell-phone is inexpensive enough for average consumers to *buy* (because of promotions that subsidize the retail price of the phone as well as cost reductions in manufacturing), but the service is still expensive for many to *use*. This phenomenon is the major cause of the reduction in the average monthly revenue per wireless customer, from between $95 and $100 to less than $40 since the inception of wireless service. Existing users were not reducing their usage in a slow economy between 1989 and 1993 and then increasing it again in the improved economy of the late 1990s. Rather, the core of users who have a high need for wireless service is diluted with customers from market segments with an increasingly marginal need for it. The marginal consumer users themselves have a bill closer to $30 or less per month.

This dilution is an important aspect of segmentation for strategy development. If newer customers from different segments are found to be price-sensitive, for example, it is important to design price strategies that appeal to these segments without lowering prices to the existing base of customers. The same applies to strategies for other segments and elements of the marketing mix.

Buyers versus Users

It is also important to distinguish the characteristics of users from those of buyers. While the profile of the highest-potential wireless buyers still holds, most of them have long since bought and retained wireless service in major markets. This segment has been saturated, and broader, different segments are now addressed. Even though less than half of users may be female, this represents a weighted average of an increasingly penetrated segment, and the current sales rate for the female segment is 50% or higher. Thus it is important to understand the makeup of the base, for both marketing and churn reduction, but it may not be a timely indicator of the characteristics of today's prime prospect in a fast-changing market like wireless communication.

The latest view of the target market points to a consumer who is between 24 and 35, is a young professional, has an income of $40,000 to $70,000 per year, is just as likely to be female as male, uses the phone for security and personal calls, and is not a heavy user. This not only makes the marketing approach different from that for a business customer, but points to different sales channels as well. This segment is not the *best* new sales target but the largest identifiable market. The best single type of customer is still the executive of a small company.

While the highest-potential segment for selling wireless services may be the executive of a very small company, the best *user* may be the middle manager of a large company. Fortune 500 companies that have adopted wireless on an institutional basis have the heaviest and the longest-lived users. *Institutional basis* means that the company has adopted the concept of wireless as a productivity tool and pays for the telephone and the service for executives in the organization according to function and management level rather than individual need. Such users are not price-sensitive (since the company pays the bill), are long-lived because managers at such companies stay with the company longer, and thus constitute a market segment with a much better-than-average usage per month and subscriber life. Once such an account is contracted, new users are added automatically. Thus, the average acquisition costs are also lower for this type of customer, making them more profitable. Such corporate customers are often referred to as *major accounts*, based on the way the account does business versus the behavior of the individual corporate user, who may never see a salesperson.

Major accounts usually buy from a carrier just because they assume the best service will come from the carrier as a principal provider. However, they will often issue a request for proposal (RFP) and may get a better price from a reseller. They cannot get a better price from a dealer, who can only offer the carrier's retail prices. They often do not buy on a national or regional basis (it is rarely offered to them because only AT&T Wireless Services and a few other providers have a coordinated national accounts program). They are interested in service features such as billing and pricing discounts, described later.

While the target markets for wireless services have changed over time, careful consideration of market segmentation and its implications for pricing, promotion, sales channels, and other aspects of the marketing mix are key success factors in wireless marketing.

PRODUCT ATTRIBUTES

Wireless carriers and resellers would like to believe that the coverage, reliability, customer service, billing, and features of their service are the prime reason for the adoption of wireless services and the selection of the carrier. This, however, is not the case. While some of these elements may influence some individual customers, and promotions of free airtime or other incentives may sometimes affect the choice of carrier, in large measure *the wireless customer focuses their purchase decision on which handset to buy.*

Whereas the wireless carrier rightly believes that the service decision is more important than the telephone, customer behavior continues to show that the intangibility of the service and the physical presence of the telephone push the decision toward the latter; the customer then subscribes to whichever service the vendor of the telephone promotes. The fact that the customer subordinates their carrier decision to their telephone decision is one of the most important aspects of promotional advertising and sales channel decisions, as we will investigate.

TELEPHONES

Wireless telephones, often called "cellphones" even though they may be cellular, PCS, or another standard, can be divided into three categories: mobile (car-installed full-power telephones; see Fig. 2.1), transportable (full-power large portable phones; see Fig. 2.2), and portable or handset (small- or pocket-sized, reduced-power portable phones; see Fig. 2.3). Originally priced in 1983 to 1984 at about $2,600, the standard cellular mobile phone had a

FIGURE 2.1   Installed wireless telephone, or "mobile" phone. (Courtesy of Nokia).

FIGURE 2.2   A "transportable" wireless telephone (Courtesy of Motorola, Inc.).

retail price of about $600 by 1991, and today can be found with a minimum of effort installed at under $300. In 1985, when mobile phone prices had been reduced to about $1,800, Motorola introduced the first real portable phone at about $3,000, and its various improved versions have become such a standard item that it has been nicknamed "the brick" because of its size and shape; today it is considered too big for many applications but is still favored for rugged outdoor and construction use. Now, virtually all standard-feature wireless phones are small portable sets that have a subsidized price to consumers of less than $100 and often close to zero if they are willing to sign a one-year service agreement.

The mobile installed and transportable phones have given way to the miniature pocket-sized portable phone. This trend has paralleled consumer adoption of wireless technology. The only disadvantage of portable phones are their lower power output. However, since most of the "holes" in wireless systems have been "filled in" with more cell sites, portable phones now

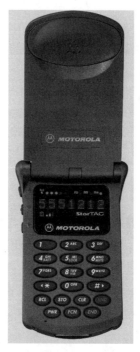

FIGURE 2.3   A true "pocket" portable wireless telephone (Courtesy of Motorola, Inc.).

provide satisfactory performance in most parts of metropolitan areas. Car-installed phones, power boosters for portable phones, and transportable phones, are only purchased when reception in fringe areas is a major consideration. Combined with price decreases, the sales rate of portable versus mobile and transportable phones has gone from about 15% in 1988 to 70% or more of all wireless telephone sales in 1992, to 98% today.

The implications of the increased penetration of portable phones are that the role of technical help in installation and maintenance is reduced and that wireless phones can be sold as a self-service item in department stores and other retail outlets without trained salespeople or technical support. This is not the recommended type of sale, but the environment that the consumer and the retailer prefer. The evolution of the mass market for wireless service has paralleled the availability of the pocket portable phone. This has implications for the sales channel organization, but also means that the mechanics of the sale are simpler and the consumer market has become more strongly attracted. The downside of the consumer market is that the consumer is very price-sensitive to the service and has a very low average usage. At the opposite end of the spectrum of market segments from the corporate user, it has the lowest usage and the shortest customer life of all of the market segments in wireless service.

Services versus Products

In this book the terms *product* and *service* are interchangeable when discussing what is delivered to the customer at a price. But there are very important differences in the characteristics of subscriber services versus manufactured products in marketing strategies and in how the efforts of the marketing process become revenues on the financial statement. As the United States has become a nation with more of its gross national product in services than in manufacturing, it is still difficult to detach marketing from its manufacturing orientation in academic literature.

In a product-based company, sales of product, rather than service subscription, produce revenue. The sale of the product is a direct result of the marketing and sales expense incurred in that period. In a subscriber service such as wireless, the expense associated with a sale is recovered in subscriber revenue over the entire period during which the customer continues to subscribe—the *customer life* of the product. Unlike products, there is no relationship between this month's sales cost and this month's revenue. This month's sales expense produces future revenue over several years (see Chap. 11, "Cellular Sales Productivity").

Therefore, a wireless carrier's purposeful reduction in sales expenses, as a short term measure to increase net income, will lower gross sales for the month, but will appear to have little immediate effect on revenue, perhaps leading to further reductions in sales budgets. However, these measures actually cause a long-term slowdown of revenue growth, which is much more difficult to reverse once it has been allowed to occur.

Another important difference in the marketing of subscriber services versus products is that products are sold just once for the full price agreed. Subscriber services, on the other hand, provide the customer the ability every month to consider the value received for the price paid and to make the purchase decision or cancel service. Therefore wireless service providers get the chance every month to either satisfy customers or face losing them, depending on their attention to system performance and dedication to customer service.

SERVICE FEATURES

Wireless service, as an intangible, has most of its features built into its pricing schedule. In most cases, the pricing is separated into service initiation, monthly access, airtime usage, and long-distance tolls. Some of the features of wireless service that distinguish it from other telecommunications services have been designed to appeal to "people in motion," as wireless users are often characterized.

1. *Local Toll and Message Charges*. Such charges are often included in the airtime charge. This distinguishes wireless service from regular, "landline" telephone service, and is an important feature of wireless service. It avoids the uncertainty of what local toll charges might be involved in a particular call and permits users to estimate their monthly bill based only on time used. It simplifies the bill by avoiding local toll detail. But most importantly, it positions wireless users' mobile status as "master of the metropolitan area" rather than being associated with a fixed location: it recognizes their mobility.

2. *Airtime*. The wireless user pays for airtime whether placing a call or receiving one. While this can be positioned as the price wireless users must pay for being available, it is really an expedient that recognizes that local wireline telephone companies do not want to bill their landline customers for airtime when calling wireless users. They have no facility to do so and do not care to have one. While some experiments with "calling party pays" (CPP) programs have been successful, it will take some time to recognize that wireless service is no longer a premium service for the privileged and should have options to be billed to the calling party.

Under the current limitations, wireless users are hesitant to give out their telephone number except to select parties and valued contacts. Incoming call completions to wireless phones are severely limited. The possibility of wireless telephone directories will remain unfeasible until "calling party pays" programs are widely accepted.

Custom Calling Features

Custom calling features are not as useful to wireless users as they are to landline services, because most wireless calls (about 80%) are outgoing. Because the wireless phone is off or unattended so much of the time, some form of answering or screening is a necessity for those users who do receive incoming calls. Since wireless users usually have to pay for the airtime on incoming calls, they give their wireless number to few people, treating it as an unlisted number, and use their office phone, secretary, voice messaging, or pager to screen incoming calls rather than answer them. Therefore, one very useful custom calling feature is call forwarding, which allows users to redirect incoming calls to these other services.

Such features are often "bundled" with premium pricing plans because they provide little revenue and cannot be economically sold separately. When they are separated, sales of these services are usually economical only with a carefully planned telemarketing program. It is embarrassing that many carriers still carry speed dialing as a feature just because it is included in system software, even though virtually every wireless phone has memory dialing. The last time somebody bought this feature was in 1985.

Voice Messaging

Voice messaging is an example of a feature that ties users to the service and performs an important adjunct service. The primary application is for *telephone answering* for users who receive telephone calls. *Voice mail*, another application of voice messaging, is the exchange of messages among people associated with each other, for example, managers of the same company, rather than live conversation. It is not a primary application in wireless service as in the office environment because wireless users do not have a "community of interest," a joint need to communicate among themselves as managers of a single company do. Because of the benefits of voice messaging in increasing the utility of wireless and providing additional revenue per user, most wireless carriers have adopted this service.

While some carriers claim very high penetration of voice messaging, a 30% to 40% penetration results only when the service is bundled with premium access plans and/or promoted as a 60- to 90-day free trial to new customers. Because of the limited amount of incoming call traffic and the number of people with existing voice messaging services at their home or business, it is difficult to raise penetration above 12% to 15%. This is still an appreciable penetration, and it is an important service to keep the loyalty of heavy users. It increases monthly subscription revenue by $5 to $10 and has been shown to stimulate usage as well as increase customer satisfaction and loyalty.

Billing

Billing is often considered an accounting or operations function, but it also is an integral part of the product. A bill that is easy to understand is a big benefit to the individual user. The provision of custom billing services to multiuser corporate customers is a competitive advantage in keeping the account, by permitting easy accounting by department, by providing the bill in tape or floppy-disk format for analysis, etc.

In combination with pricing plans, billing can be a product definition in itself. Prepaid wireless service, for example, is a product defined entirely by the way the service is billed and its pricing plan. Even simpler, special billing treatment of services such as roaming and long distance can constitute products without any other modification of the service.

Other Product Elements

Additional elements of the product dimension include the development of products such as wireless data transmission and other wireless-based services for special markets. While these markets are slow to emerge, they will play an increasing role in wireless services as penetration of additional high-payoff markets is attempted. Chapter 15 deals with new types of services.

FURTHER MARKET SEGMENTATION

A subdivision of the segmentation of wireless users can be termed *early adopters*. As we look at the history of wireless service, the initial primary market segment was the set of executives of small businesses and professionals (lawyers, doctors, etc.). The next segment to adopt wireless services was business managers of larger organizations, especially in the target industries (real estate, construction, and financial services). Third, large corporations adopted wireless technology on a policy basis, that is, companies adopted and paid for the service and phones for classes of managers, determined by either level of management or functional areas that require extensive field work. Finally, the consumer segment emerged in the early 1990s as the largest segment. Each of these segments emerged as wireless phones and service became more available and less expensive. Not only did they add marginally to the size of the market, but each new segment was larger than the previous ones (see Fig. 2.4).

As each new target market accepts wireless services, additional dimensions to the segmentation of wireless users are inserted based on how users make the decision to employ the service. Thus within each segment are the early adopters. These are the people who are comfortable with technological change and are anxious to have the latest electronic devices. The nurturing of the early adopters is important, because these users provide word-of-mouth credibility to new technologies and informally shape the determination of the real value of the technology as either a "fad" or as a long-term provider of benefits and solutions to problems.

The early adopters of wireless communications confirmed that the technology was sound, that the telephones were quality units, and that they

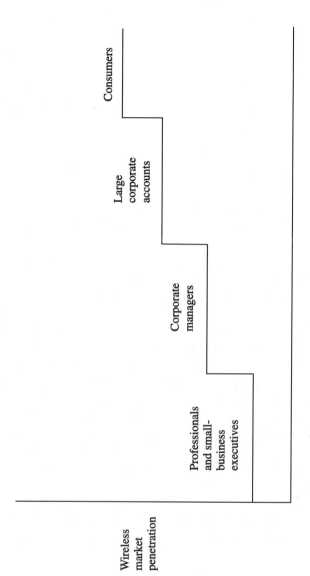

FIGURE 2.4   Wireless market segment penetration.

found true value in productivity gains and communications availability. More important than even these considerations, wireless avoided the pitfall in which most new products fail: it was easy to use. It did not require education and user training, and the concept was easy to grasp. It provided users the opportunity to do things better without asking them to adapt to it. Early adopters in the consumer segment have not always found the same value as business customers. They use wireless service less and are more sensitive to price than early adopters in business.

These characteristics of the developing wireless market lead us to an important understanding: newer markets must go through the progressive stages of user development within these market segments, and the confirmation of the service value among early adopters. Thus younger, newer wireless applications providers may not immediately "jump" into an environment of addressing a consumer market through retail channels with the latest in small portable telephones; each market must go through the stages of technical and marketing maturity, although in newer markets it may occur more rapidly.

These stages include the technical development of the system as well as marketing development. Portable telephones, for example, originally did not operate well inside buildings and in "fringe" areas in newer markets. Initial users of a cellular system had to pay careful attention to the installation of a mobile phone and the orientation and type of antenna used on the automobile. As the wireless system built a user base and generated more and more usage, more cell sites were required to be spaced at closer intervals. These additional cell sites, together with careful power adjustment, channel coordination, and antenna propagation, began to provide a fairly uniform signal strength, and transportable systems with lower and less carefully oriented antennas provided adequate performance.

Finally, with full system deployment, users found adequate performance with low-power portable phones inside buildings and in fringe areas, at any orientation of the antenna. Early users of portable phones had often used them in an awkward posture in order to maintain the portable's antenna in a stationary, perfectly vertical position.

Newer features such as voice dialing and data applications face the problem of maturing technically while the user matures in his or her ability to translate product features into their own benefits and solutions and to use new technology effectively.

3

THE MARKETING ORGANIZATION

THE EVOLUTION OF SALES

The argument concerning the overall control of revenue production between marketing and sales organizations has never been settled. In most organizations that use a direct sales force, the sales department controls revenue and the marketing department is a support arm.

For many consumer products, however, marketing controls revenue, often merely by default because there is no company sales channel. Independent retail stores, the major sales channel members, are not under the control of the companies that produce the products they sell. In these organizations, the product manager may be responsible for the entire income statement for the product division as well as revenue.

Direct sales forces, or any sales mechanisms in which the manufacturer or service provider is in control of the sales channels and sales process, usually exist if the company (1) has several or several hundred products aimed for a narrowly defined market; (2) has a product or product line that has a high price (a "big-ticket" item) and/or technical orientation deserving the cost of a direct sales force or other dedicated channel, such as a national chain of specialty stores; (3) has an industrial customer base that prefers to do business with a sales representative in their own office; or (4) has a mixture of these.

Wireless industry managers recognized early that the first wave of core customers would be mostly business people who would require a visit from a professional salesperson to understand the benefits its of wireless service. The sale would require not only wireless subscription but the purchase of a $2,600 phone and technical expertise in installing the system correctly and with care, respecting the high-end motor vehicles in which it would become a permanent fixture.

Thus wireless systems originally had all the elements that indicated the need for a direct sales force. Some carriers developed an internal sales force and installation capability. However, most found benefit in working with contracted sales agents and dealers who would combine sales efforts of wireless service and wireless telephones with professional telephone installation, technical support, and customer service in a single, local, one-stop shopping facility. The Bell mobile telephone companies were initially prohibited from offering phones through the same division as telecommunications services anyway.

As wireless technology has matured, many of these requirements of sales channels have been modified with changes in technology, customers, and other influences. As mentioned earlier, the average retail price of phones decreased from $2,600 to under $600 in the first few years. Therefore it became difficult to justify a direct sales force that was paid high commissions in return for telephone profits. Not only have prices continued to decline, but competition among equipment manufacturers combined with competition for customers has driven the retail profit out of the wireless phone business.

Increased customer awareness makes it less necessary to convince prospective customers of the benefits of wireless service. Many prospects in the consumer and small business segments are more comfortable buying through retail channels than from a direct sales force. These potential clients self-select themselves as sales prospects rather than require qualification as "leads" for a direct sales channel salesperson.

Finally, the increased penetration of low-power pocket portable and transportable phones and the availability of quality car installations have reduced the importance of installation as a concern. Further, improvements in the way wireless phones are programmed, are tested, undergo troubleshooting, and are maintained, and improvements in their operation because of improved wireless systems and deployment, have diluted the importance of technical support.

Thus wireless sales channels no longer require a direct sales force nor a technical capability to sell and service market segments. Wireless service is sold through many channels in which it is not the only product or service provided, such as automobile dealerships and consumer electronics stores. Like other products that are sold through multiple channels, the sales organization is no longer in control of many sales, and marketing has an increased role in "pulling" customers to channel outlets, rather than sales "pushing" salespeople to qualified prospects.

Any marketing organization that is controlled by the sales responsibility of a direct sales force will be driven by the sales department, and the marketing department will usually end up as part of the sales organization. Unfortunately, many sales organizations downplay the value of marketing. Often, the marketing department will be a subordinate function of sales, which is understaffed, unsophisticated, little understood, and given little autonomous responsibility. With that the sales organization's classic portrait of the marketing department as a do-nothing staff organization becomes a self-fulfilling prophecy.

Increased Role of Marketing

As the diversity of channels and market segments has increased in wireless telecommunications, the marketing function has a larger role in presenting the service to potential subscribers, and the sales function plays a diminished role outside of retail and direct sales. At some point, the ability to understand market segments, the need to match channels to segments, and the need to use "advertising pull" instead of "sales push" as the predominant means of putting the prospective customer in a purchase mode takes over and makes marketing the dominant, rather than the subordinate, function in the overall marketing–sales cycle.

Advertising pull refers to the use of an advertising offer to bring customers who are far along in the marketing awareness stages to a purchase decision or sales situation. Sales push refers to the need for a direct salesperson to find prospects, nurture leads, and work to bring the customer to a purchase decision through the sales presentation.

In most wireless carrier organizations with some experience in the marketplace, the management of the direct sales function needs to be separate from the management of the broader external sales channels, and the overall marketing function should be separate from both. The marketing function should be on an equal footing with the overall sales function, if not supervisory to it. The skill set of marketing is entirely different from sales. Marketing must not be a part-time pursuit of the sales organization (see Fig. 3.1).

The Marketing Function

The marketing function is divided into several areas under the general headings of product management and market management. The mix of

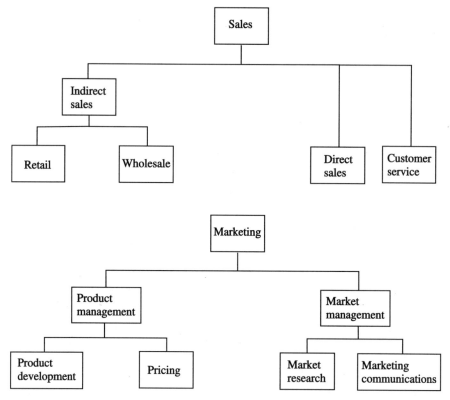

FIGURE 3.1   Sales and marketing functional organization.

functions under these headings, as well as the subordination of one to the other, is an open issue.

Generally, the function of market management is that it acts as the surrogate representative of the customer or the communications conduit with the marketplace, embracing market research, market segmentation, and marketing communications, which include advertising, promotion, and marketing-oriented public relations.

The product management function is associated with characteristics of the product that satisfy the needs of the market: the design and content of the service, its pricing, and its terms.

In the context of the four P's of marketing [product, pricing, promotion, and distribution (physical delivery and sales)] product management is comprised of the product and pricing elements, while market management controls the promotional element in combination with the understanding of the marketplace in terms of market segmentation and market research. Distribution, the fourth "P", is separated into physical delivery, in the case of a manufactured product, and sales. In wireless service, the physical distribution function in its purest sense is limited to the role of customer service in delivering service initiation, or *activation*, and supporting wireless as an intangible service. Since the sale of a new phone is usually entailed, however, the product delivery function usually remains. Organizationally, sales and customer service usually report separately to a general manager or marketing and sales director, because of the size and span of control each function requires.

In a product or service for which there are only one or few products going to many markets, market management dominates the organization and usually controls its direction; sales are determined by how well diverse markets are managed. If there are many products and few key market segments, product management dominates; sales are determined by how well each product fits the needs of one or a few key customer segments.

In the best wireless organizations, the two managements cooperate rather than compete for organizational superiority. Strong personnel in either capacity can also be a deciding factor in who takes the lead. Early dominance of the product management function in designing all aspects of the wireless offering has yielded in the maturity of wireless services to the dominance of the market management function as subsegmentation, new market penetration, and promotional design have gained increasing importance.

ORGANIZATION PRINCIPLES

A marketing-driven organization considers that sales channel management is just one part of marketing. Thus, marketing functions direct the sales effort. In more conservative organizations, marketing has been merely a sales support function. With the advent of sophisticated marketing techniques and multiple sales channels, marketing becomes the driver.

The organizational implication is that marketing and sales should be a combined director-level responsibility or that sales reports to marketing. This is in contrast to organizations in which sales and marketing are completely separate responsibilities reporting to the president or chief operating officer, or where marketing is subordinate to the sales function.

In the market-driven company, the marketing forecast is based on market factors and market research and is the basis of the business plan for the company and the common forecast for all departments, although marketing's forecast is usually significantly revised to reach consensus by senior managers in all functions. Even though marketing produces the annual sales forecast for agreement by management, marketing is a staff organization. Sales management takes line responsibility for achieving the revenue forecast when it consents to the annual forecast. The sales area often sets incentive-based sales goals higher than the business plan forecast but still operates nominally at the planned level.

Within the marketing function, the organization is usually formed as with the dominant function, around a product orientation or a market orientation. Product orientation is dominant if there are multiple major products going to a narrow set of customer segments, or if each product is mainly targeted to a defined market segment. In this case the linkage between the product and its application is apparent.

Market orientation, or organizing marketing according to customer segmentation, is dominant if there are one or more products each going to multiple segments with different needs, channels, and characteristics. In this case, the important element is packaging the application of the product for delivery to different audiences.

If several products are targeted for the same few segments and there is not too much difference in the mix of segments for each product, either product or market management can dominate, and it often depends on the need for expertise in one area of product- or market-based functions.

Because mobile communications has a small number of products all going to several very different segments, a market management emphasis is preferred. Within the marketing organization, processes are designed around market segments, with segment managers as specialists in under-standing and designing segment-based marketing programs. The forecast is based on market segments and channels, and this division of the forecast is more important than a product orientation for companies in which the core business is wireless services.

As previously mentioned, the marketing organization is generally sepa-rated into product management, which includes managing existing products by forecasting and designing changes and additions to the product, product development, and pricing. Market management includes functions that are considered to be a surrogate for the voice of the customer—market research, segmentation, marketing communications and advertising, and customer communications (bill inserts and customer letters, telemarketing and customer service scripts, marketing press releases, etc.)

Additional functions in marketing include forecasting and business development. Forecasting may be in the marketing management or market research area, or may be assigned to a separate manager reporting to the marketing director who does all modeling and budgeting. Business development involves finding major alliances, major new product areas, and major customers and is not a common function in the marketing organization of a cellular carrier.

The sales function in a multichannel company is more commonly called sales channel management and includes channel managers for each channel.

The channel managers are responsible for achieving the sales forecast for the channel, and this includes the responsibility for adding channel members as well as supporting and developing existing channel members. Adding channels is the responsibility of the channel manager as supported by the requirements of increased market share and the channel preferences of segments recommended by market management. Such functions as sales support—making sure that sales channels employees have the tools needed to accomplish their job—may separately report to the head of sales channel management or the individual channel manager.

The typical marketing and sales organization is charted in Fig. 3.2 and Table 3.1. Not all levels of management are necessary if the span of control is less than seven or eight subordinates between levels.

Alternative 1 is the most modern and sophisticated marketing or sales oganization. A single vice president has line responsibility for revenue and sales and uses staff marketing resources to define most of the elements of marketing, with the sales department as the line organization actually delivering the sale. Customer service, as a line organization that interfaces with consumers, reports to the sales department, but has a marketing-driven orientation rather than a billing inquiry and/or service suspension orientation.

The organization shown is not structured the same way in a recommended decentralized environment. All of the functions exist, but most report to an area general manager. Headquarters functions include national accounts sales, marketing research, segment managers, advertising and pricing planning (advertising implementation and price plan development are done regionally with HQ approval).

PROCESSES AND RESPONSIBILITIES—PRIMARY ORGANIZATION

Vice President of Marketing and Sales

- Line responsibility for sales and staff responsibility for all marketing functions

Director of Marketing

- Staff responsibility for all marketing tasks except sales

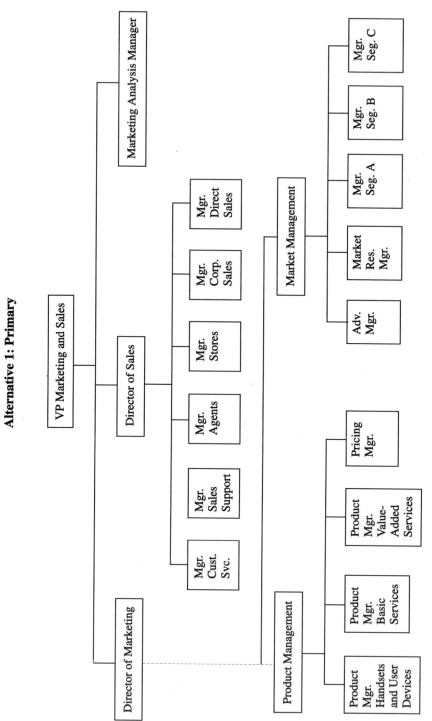

FIGURE 3.2 Typical marketing and sales organization.

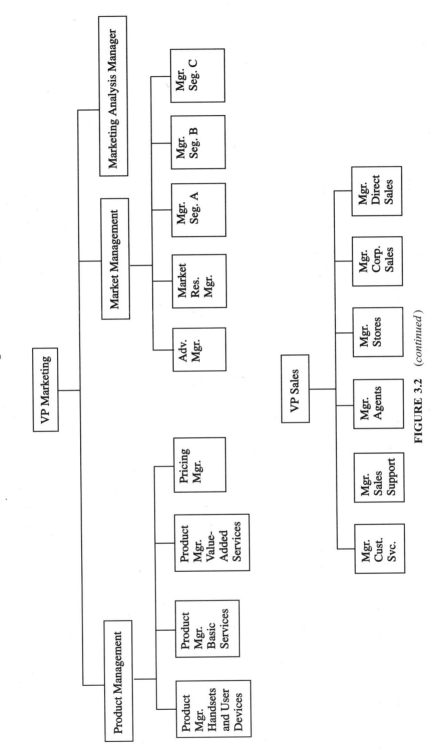

Alternative 2: Separate Marketing and Sales

FIGURE 3.2 *(continued)*

Alternative 3: Sales-Driven

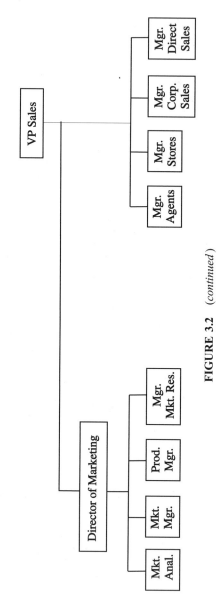

FIGURE 3.2   (*continued*)

TABLE 3.1   Organization Advantages and Disadvantages

Alternative 1
Advantages   *Disadvantages*
Marketing group drives sales   Staff overhead
group to best segments   Decision making takes time and
Information used most effectively   study
for increased sales and revenue   Segmentation and analysis
Sales department spends more   systems and procedures required
time selling

Alternative 2
Same as above   Slow communication and peer
                                      interaction between marketing
                                      and sales groups
                                      Sales department may not take
                                      direction from marketing
                                      manager
                                      Sales department may develop
                                      their own marketing capability
                                      when at odds with marketing
                                      department

Alternative 3
Lean organization   No intelligence used in selling
Quick sales decision making   Market research and competitive
                                      analysis are mostly sales
                                      anecdotes
                                      Poor integration with other
                                      functions
                                      Little attention to planning

Product Management

· Product management
· Product development
· Pricing

Subordinates

Product Manager

· Product management: product volume, product design, product variations, features, marketing specification, and user manual and literature content

- Product development: specification of new products, vendor application and product evaluation, product management to commercial deployment [trial, vendor management, research and development (R&D) management, etc.]

Pricing Manager

- Pricing plan development
- Pricing plan analysis
- Terms and conditions (length of contracts and penalties, bundling, quantity discounts, etc.)

MARKET MANAGEMENT

- Market segmentation
- Marketing communications and advertising
- Market research

Subordinates

Advertising Manager (Marketing Communications Manager)

- Chooses and directs advertising agency or directs in-house agency
- Designs advertising and promotions with input from product management, market management, and sales management
- Tracks advertising results
- Plans media, media frequency, media integration, and places media
- Publishes marketing and product literature
- Produces marketing press releases
- Produces marketing events, public relations interviews, etc.

Market Research Manager

- Conducts secondary research (demographics, competitive analysis, industry analysis and news)
- Conducts primary research (focus groups, advertising effectiveness, customer satisfaction)
- Manages market research and customer data

Market Manager or Segment Manager(s)

- Maintains information on segment or segments for users and nonusers
- Establishes and manages segment forecasts
- Researches segment needs and applications
- Designs segment-specific advertising, promotion, and public relations concepts

Market Analysis Manager

- Marketing forecast from market research and other inputs
- Financial analysis of marketing and sales results and deviations from forecast
- Budgets for the marketing and sales area
- Models pricing and more

Sales Director

- Manages and develops all sales and channel functions
- Has line responsibility for achieving the revenue forecast

Subordinates

Channel Manager

- Manages, trains, and motivates channel and develops new channel sites and/or members
- Has line responsibility for channel sales results

Sales Support Manager

- Supervises sales training
- Designs sales incentive programs, sales meetings, and sales awards
- Assures delivery of marketing and merchandising materials to sales channels
- Develops sales handbook, sales presentations, sales aids, and sales automation tools
- Tracks sales
- Generates leads for sales

Customer Service Manager

- Manages and trains customer service representatives
- Maintains service standards for customer service Key Performance Indicators (KPI's)

Of, course, since most carriers organize wireless service by geographic market, the sales or marketing organization may have only one manager. Members of the organization, regardless of its size, must understand the various functions as they relate to the marketing process. Members must also respect the independence and objectivity required when considering the needs and markets of retail versus wholesale channels in the sales department, or market research versus product development in the marketing department. The local general manager should at least be able to separate the sales and marketing functions by appointing two different managers. In larger markets, the sales function becomes just one element of the overall marketing function under the direction of the senior marketing executive.

Newer markets are primarily driven by the sales function. As markets grow and demand a more sophisticated marketing approach, the sales function becomes a key subordinate of the overall marketing organization and function. Some organizations both in and out of the wireless industry, however, never mature to acquire an understanding of the importance of the larger view of marketing. Marketing is then retained as a subordinate and peripheral function of the sales organziation, often viewed as providing only marginal contribution. This view can often be the prime cause of a marketing organization's poor performance.

DECENTRALIZATION

Wireless carriers that have operations in more than a few markets may be organized on a centralized or decentralized basis. Decentralized markets are run by a local general manager who has control of most resources and functions for the marketplace. With centralization of functions such as human resources or billing, the general manager has an influence on the central administration of the function and assumes an allocated portion of the expense.

There are two major advantages with this type of organization: speed and responsibility. The general manager (GM) has complete profit and loss statement (P&L) accountability for the financial performance of his or her unit and should have the freedom of responsibility to make decisions and take action to control the unit in most areas. Second, the decentralized organization can assess a local problem or opportunity and take immediate steps to correct or take advantage of it. One of the prime characteristics of the wireless marketplace is that the carrier has different competitors in each market, increasing the need for a local marketing orientation.

Often, the prime area of concentration for any market is marketing and sales. Accounting, operations, and engineering departments provide great opportunities for service and efficiency improvements, but are fairly straight-forward within the capital budget and expense constraints. Marketing and sales provide the greatest variability in terms of opportunities and constraints, or revenues and costs, over a wide range of variables and competitive conditions, requiring creativity and skill. Marketing and sales managers are often recruited for general managers, and the job is about 70% marketing and sales. Because of the P&L responsibilities, financial skills are also a prerequisite for the position, enabling the GM to glean opportunities and outline corrective actions from the analysis of numeric performance data.

The prime disadvantage of a decentralized organization is the lack of skill depth in any particular discipline. Functions such as advertising, public relations, promotion development, and pricing are very shallow and are often left to the sales organization as a sideline. These skills become more important as initial core market segments are penetrated. The need for careful market planning and competitive marketing programs increases with the increasing scarcity of pent-up demand, core market opportunities, ready buyers, and low marketing costs.

CENTRALIZATION

In a centralized organization, each function (marketing and/or sales, finance, operations, engineering) in a local market reports to a superior in the function at headquarters, and the P&L is consolidated to the company level. Each function has a budget and multimarket performance objectives.

Such an organization has depth of expertise in each functional area, enabling it to create and make use of marketing intelligence with programs that do an excellent job of developing markets and providing superior marketing communications materials, sales training, etc. But any advantages in such areas are severely constrained by disadvantages. Without any local orientation, marketing programs tailored to a specific market are never available when they are needed; when available, they are not sensitive to specific, local market needs. Productivity is lost in constantly trying to devote the resources of 10 marketing experts to 10 markets 10% of the time. Most painful is that when one or two of the required resources are available from headquarters for an opportunity in a specific market, the other resources needed to complete a marketing program are not available. Coupled with the situation in which none of the marketing people is accountable to a local market organization, the problem moves from unwieldy to impossible to solve.

While no single solution to the problem is available, it appears that the most effective solution is to have a decentralized organization with marketing depth in the headquarters organization. Members of the central organization serve in the following ways:

1. Develop market intelligence and conduct market research that is specific to each market and consistent among all of them.
2. Assist local markets in developing local expertise and regular promotional programs, and review proposed local programs for adherence to standards, such as use of logos and presentation of the company.
3. Create promotions and sales support materials of professional quality that are easily adaptable to local markets.
4. Devise and/or support local marketing programs as pilot programs and new market development efforts before considering broader deployment.
5. Take a successful, home-grown marketing program from a local market and package it for additional markets.
6. Investigate new products, applications, and markets, such as wireless data, which local markets cannot do effectively.

In such an organization, annual marketing plans are developed by local markets based on assumptions supported by headquarters (HQ) marketing staff. Local markets require bottoms-up market plan development in order to set their own priorities, develop plans in the context of their overall projected financial performance, and commit to delivering it. The pitfalls of such an organization are in the following categories of problems:

1. The HQ marketing organization is used as an auditor or reviewer of local marketing programs rather than just for adherence to standards.
2. Local markets resist marketing programs that have been successful in other markets in favor of home-grown programs that are less sophisticated but are "bought into" by local marketing and sales managers.
3. The HQ marketing function produces program recommendations that are not responsive to individual markets nor ready at the time individual markets need them. Where they become available, local markets have no capability to convert them into specific programs.

4

THE STRATEGIC PLAN AND THE MARKETING PLAN

To managers in many industries and disciplines, planning is merely an activity for idle hands. Poor planning with poor results and little follow-through produces poor attitudes about the planning function and a vicious cycle of even less attention paid to the process. Some argue that plans always change drastically within weeks of their drafting, so why even bother?

For marketing programs to be consistent and rational, all the elements of the 4 P's must work in concert. The programs must be part of larger growth and financial goals and must be part of a larger corporate strategy encompassing all disciplines in the corporation. It is important to have consistent marketing plans as a product of a good corporate planning process. However, marketing also provides the primary quantitative input to the corporate plan: the revenue forecast based on market sizing.

Understanding that plans are made to be changed, there are several important reasons why strategic plan development is an important part of wireless marketing:

1. A well-constructed plan provides the basis for corporate and management performance measurement.
2. It includes an annual plan that provides a benchmark for results and guidelines for corrective action, based on the degree of deviation of actual results from the plan. It also provides an analysis of the causes of deviations: market and cost variations, poor performance or incorrect strategies, for example.

3. The development of a strategic plan is of value at least as much for its process as the resulting document. No other process brings corporate officers and managers together to develop basic assumptions and reach a consensus on a joint focus as much as a team effort to develop the plan.

We will endeavor to outline the formation of the corporate mission, strategic plans, and the marketing portion of the annual operating plan in the context of wireless services, using hypothetical examples. We will draw on these to show how top-down strategies fit with bottom-up intelligence to provide a coherent, integrated plan for wireless marketing.

THE STRATEGIC PLAN

Ideally, the development of actual marketing and sales programs, advertising strategies, pricing structure, levels, and terms, etc., is part of a much broader strategic process. First, the corporate mission, strategic plan and objectives, and marketing plan must be laid out and agreed to by top management. These form the basis of the marketing mix strategy and the pricing plan. The quality of the strategies at each succeeding lower detail level will be a function of the diligence applied to the higher-level strategies on which they are based. The pricing strategy and market forecast are closely intertwined with marketing's quantitative goals (revenue) and the financial goals—net income, cash flow, or other profit measures—and are based on these important marketing inputs. The lower-level marketing and sales strategies and tactics confirm the company's ability to reach these goals. Therefore, the overall financial objectives, as well as the marketing strategy, are key to developing a strategic view of each of the marketing functions. The process is iterative. The development of the final marketing plans may require the adjustment of the revenue forecast at the root of the plan. Modified forecasts may change strategies in each area of marketing and their relationship to each other, as well as change strategies in other functional areas.

THE CORPORATE MISSION AND OBJECTIVES

A wireless carrier might first state its corporate mission as "to become the leading provider of wireless service in the region" and state its objectives in terms of growth and profitability. However, such platitudes provide no direction to employees nor the subordinate strategies that flow from it. "Leading provider" is uninformative. Leadership must be defined in terms of size, profits, innovative products and promotions, adoption of new technologies, and positions on influential industry forums.

Growth, for example, must be defined in terms of size, character, and timing. A specific growth goal for a startup carrier might be to achieve a level of 1,000,000 minutes of usage per day within three years. This goal would avoid the problem of concentrating on customer growth at any cost and perhaps of attracting customers with low usage characteristics at high sales costs. But this growth would also need to be defined in terms of customer growth at an average usage level.

The most crucial test of a strategy or objective is to determine if any different behavior is implied when the strategy is stated in the negative. Thus if the behavior implied by *not* becoming the leading wireless carrier remains the same, a more specific definition of the goal is required.

It is important to distinguish the difference between strategies and goals. In our terminology, *goals* are quantitative results to achieve; *strategies* are ways to get there. These two terms have been confused even in textbooks. If a goal entails reaching a specific customer count by a certain time, it should state the market segments to which they belong, the usage they should generate, the mix of peak and off-peak traffic, etc. These details will help you determine specific strategies that will get you there.

A goal to reach a certain penetration of the corporate market segment might define a strategy that includes a direct sales force of a certain size, an extensive training plan, an outbound telemarketing department for generation of leads, a corporate pricing plan, and flexible payment terms for slow-paying (but sure-paying) corporate accounting departments. A goal to reach a specified penetration of the consumer market might specify a strategy to use retailers as commissioned dealers, an extensive cooperative advertising plan, an inbound telemarketing operation that can close as well as qualify, a tight credit and payment policy, and a special rate plan.

The corporate goals are mostly concerned with the parameters of profit and growth; the subordinate sales goals involve revenue generation goals, but will have goals for costs also. The difference is that the corporate profit goals recognize corporate revenue and expense; marketing goals encompass all corporate revenue (for the wireless operation) but only the marketing and sales portion of cost.

SAMPLE STATEMENTS

Mission

To establish a wireless carrier operation that is grounded in the development of a service that derives its competitive advantage through superior equipment and design, superior customer service, and premium pricing to that segment of wireless customers who demand the highest service standards and perceive high, long-term value in wireless service. The company is thus focused on a long-term, major market presence using the highest-quality equipment and personnel, providing maximum financial return to the shareholders through customer loyalty.

Strategic Objectives

Maximize financial return as measured by cumulative gross operating income, that is, net revenue less operating expenses or income before interest and taxes. Cumulative measures in planning ensure prioritization of long-term objectives and growth strategies over short-term opportunities.

The company will endeavor to concentrate on growth rather than cost savings as a source of long-term financial improvement.

The company will not forego investment and system growth in order to achieve short-term return on capital investment.

Strategies

1. Develop long-term customer loyalty at premium prices.
2. Expand to additional markets through acquisition to increase growth.
3. Concentrate on a target market of highly mobile business executives as outlined in the market forecast, whose needs and willingness to pay are consistent with the premium service offering, premium pricing, and quality customer service assumptions. Develop a secondary market of upscale consumers.
4. Establish internal sales channels for premium corporate customers, high-quality external agent or dealer channels for the target market, and establish external alternate retail and dealer channels to test penetration of additional market segments quickly.

These are examples of such statements and are not intended to state unilateral strategies superior to any others. The importance of the examples is to show how strategies fit with objectives and complement each other. At this level, the actual quantitative goals are not specific nor time-dependent.

For the sake of brevity, the strategies and assumptions given in the following mainly involve pricing, but are based on a complementary relationship with other marketing disciplines. Similar strategies should also be developed for product, promotion, and distribution. As each set is developed, they should be compared for complementary relationships with the other disciplines and modified until they are cohesive, integrated, and consistent.

SAMPLE MARKETING AND PRICING STATEMENTS

Marketing Objectives and Pricing

1. Achieve a 12% penetration of the market population over three years.
2. Achieve revenues over the same period equivalent to $60 per month per customer or greater.
3. Maintain a marketing expense less than $400 per new customer.

Strategies to Achieve Marketing Objectives

1. Establish an image of quality differentiated from competing carriers through exceptional customer service, quality engineering, and premium equipment.
2. Do not use low price as a market entry vehicle; concentrate on superior value. Create a premium price strategy consistent with an image of premium quality.
3. Establish premium price levels with price structures that appeal to several major market segments, without appearing to "cream-skim" the market by pricing too high.
4. Establish retail channels that provide quality service delivery to the end user, the quality of which can be strictly controlled.
5. Establish alternative channels that penetrate price-sensitive and marginal market segments at lower marketing costs.

Business Assumptions for Pricing

1. The Company will concentrate on wireless service and will not attempt to make equipment and accessories a major business or profit center in competition with wireless service or the success of other channels.

2. The company will utilize a mix of sales channels to ensure efficient delivery of service to all types of customers, to provide opportunities for other local businesses in wireless in partnership with us, and to penetrate target market segments quickly:

   A. Direct sales force for major corporations who need long-term decision support and follow-on support

   B. Local upscale retailers that have not been significantly penetrated by the competition to attract upscale consumers

   C. Exclusive sales agents as a substitute for carrier-owned stores, dedicated exclusively to wireless service, to provide high-quality sales and service, one-stop shopping, and professional, local customer service

   D. Resellers buying at wholesale prices to provide more rapid growth in net income, lower average marketing cost, broader market penetration, and broader pricing and service options than carriers can provide

   In order to best control the quality of retail delivery, the company will target its internal, dealer, retailer, and exclusive agent channels to assume approximately 80% of the retail sales capacity of the carrier operation, assuming that resellers and independent agents can deliver the other 20%.

3. Prices may vary from one market to another depending upon demand, need for service among target segment population, size of coverage area, etc. A multimarket single rate will be available on some high-usage plans for the mobile heavy-user segment.

4. Bulk service at wholesale to resellers will always be lower than or equal to the lowest retail prices to the largest end users, to provide a large opportunity for alternative retail providers, while maintaining premium retail prices as a price "umbrella."

5. The pricing structure will be simple and targeted to attract the highest-potential market segments.

6. Retail service through internal and agent channels will be branded with the highly visible corporate logo and premium branding. The wholesale service shall carry the name of the company at the option of the reseller but not its logo. Internal channels and agents will have exclusive use of the logo for advertising, branding of equipment, and service identification.

7. Pricing will be based on customer willingness to pay, competitive rates, and perceived value, as constrained by financial return criteria, rather than target rate of return or cost-plus methodologies.

8. Carrier competitors cannot bear a financial cost to price more than 25% below us; even if price differentials are this great, our superior system and service will be competitive to the target markets.

Specific Pricing Structure Assumptions

1. Regulators will permit market-based pricing and rapid, competitive pricing moves.
2. Pricing structures that include premium services in basic rates are most attractive to the target market (e.g., local toll and access charges included in airtime; detailed billing and vertical services included in access), rather than least-inclusive pricing with all extras added separately.
3. There are no known economies of scale nor system equipment price reductions that can be used to assume further longer-term wholesale and retail price reductions even under competitive pressure.
4. Wireless service has a very low price elasticity for the target markets that have a high need. This means that there is no increase in demand when prices are lowered such that total revenue is increased. Price structures for target markets should always be priced as high as possible. Lower prices, which attempt to attract new customer segments that are more price-sensitive than core target customer segments, should always be promotional rather than permanent and should not be crafted to attract existing customers on other plans.

Complementary Marketing Strategies and Assumptions

1. Strong brand advertising will drive penetration of the target market rather than price advertising, but competitive pressure will require promotional price reductions.
2. Pricing below competition as a market entry or penetration vehicle is not appropriate, given other marketing strategies:
   A. It does not fit with a premium service image.
   B. The target market is more sensitive to coverage and quality than price.
   C. Creative advertising, promotions, and sales incentives achieve sales objectives better. Undercutting the competition on pricing will leave money on the table if these measures are effective.
   D. Incumbent carriers have always met or exceeded competitors' price reductions.

These are examples of how higher-level objectives and strategies provide the basis for more detailed ones at the tactical level. While these examples concentrated on pricing, they affect all aspects of the marketing mix and strategies for product, promotion, and distribution. We will defer actual pricing to another chapter.

These particular objectives and strategies might actually be useful to some carriers but are merely illustrations of how such strategies and objectives must be explicit, consistent among all elements of the marketing mix, and measurable. Without such explicit statements, managers often leave the room when broad, overgeneralized strategies to achieve "quality" are laid out. Managers perhaps contract sales channels without regard to service quality and market segment concentration, or produce pricing plans that do not fit the overall strategy without the guidance they need.

There is much information available today that might lead wireless business people to entirely different marketing and pricing strategies: the dominance of the consumer market, limited third-party distribution channels that are available in some markets, and other influences. The key is to ensure that all objectives and the strategies used to realize them reflect the same information and assumptions, and that product, price, promotion, and distribution strategies complement each other in the achievement of the objectives.

The next layer of such a strategic plan development would include one- and three-year goals for revenue, cumulative operating income, and customer forecasts, usage assumptions, sales force sizing, price levels, etc., that support them. Additional portions of the plan would outline assumptions, objectives, and strategies for each functional area—finance, billing, information systems, engineering, operations and maintenance, and human resources, for example. For a decentralized company, each region or market would adhere to the basic assumptions and strategies but include different objectives and forecasts (which quantitatively combine to yield the corporate projections), areas of concentration, programs, and tactics.

The three-year strategic plan differs from the one-year plan because it includes areas of development with a horizon greater than one year: the growth of a new sales channel, the cultivation of a new market segment, addition of new products, technology changeovers, etc. It has more uncertainty in its forecast and less detail in the determination of tactics and resources to accomplish its goals. However, it provides the opportunity to extrapolate beyond the limited resources and constraints of the marketing plan to take more advantage of unrealized growth potential.

The Annual Marketing Plan

The one-year plan constitutes the immediate operational plan, forecast, and budget for marketing as an integral part of the one-year operations plan for the entire business. It lays out specific programs for the coming year and must detail specific programs to achieve quantitative results.

Therefore the marketing plan must outline and relate quantitative forecasts and costs directly to the following, for example: the total forecast of new customer gain, the number from each market segment, the amount of penetration of each targeted market segment required, the advertising programs used, the media schedule and the number of sales leads generated, the number of sales in each channel, the commissions paid to each channel for each kind of customer, sales per salesperson, and gain in channel size required to achieve the forecast.

Assuming the carrier has already been in operation for a year or more, the basis of the marketing plan is the expected revenue from the existing customer base, recognizing "churn," or customer attrition. Growth in the customer base should recognize the replacement of lost customers in the current base as well as losses occurring during the year, in addition to the growth required to achieve a forecasted end-of-year increased base. While the gross sales levels required should recognize the need to replace churn, sales should not be held responsible for actual net gain (gross customer gain less churn) but for achieving their gross gain objectives. Churn cannot usually be controlled by sales channels; however, it can be manipulated. If commissions are paid for customers who last less than 90 to 180 days, the introduction of a minimum subscriber life greater than 180 days for commission payment (paid at the time of activation but credited against future commissions when a customer leaves the service) will substantially reduce short-term churn.

The total revenue from new customers must take into account the concentration of new customers in different segments from the base, weighted by the varying revenue per customer of each of these segments. The revenue from new customers must be distributed throughout the year in concert with the timing of actual customer gain, month by month.

The bulk of the forecast should be underwritten by committed quotas from internal and external channels. Detailed programs for immediate acquisition of new channels and channel members (new agents, sales force and quota increases, etc.) should be documented in the plan where gains are not assured through quotas and commitments. See Table 4.1 for a sample marketing plan forecast.

The forecast for new customer gain must be reconcilable to the known productivity of salespeople, locations, sales force size, and/or stores. For direct salespeople, the number of qualified leads required should be quanti-fied, and the advertising, promotions, and telemarketing programs and staffing required to generate them should be recognized. It should be shown that the highest-cost channels are producing customers in the high-est-usage, longest-lived customer segments.

TABLE 4.1   Excerpt from Marketing Plan Forecast

Beginning subscribers	40,000
Net gain objective—20%	8,000
Ending subscribers	48,000
Churn at 3.3% per month	19,008
Gross gain required	27,008
Average customer base	44,000
Revenue	
Startup fees at $25.00	$675,200
Monthly access (44000×12)	
5,000 @ $25.00	$1,500,000
18,000 @ $50.00	$10,800,000
21,000 @ $40.00	$10,080,000
Excess Peak Usage (44,000 × 12)	
5,000 @ 50 min × $0.80	$2,400,000
18,000 @ 80 min × $0.40	$8,448,000
21,000 @ 120 min × $0.36	$10,886,400
Excess Off-Peak Usage at 20% Peak	
5000 @ 10 min × $0.24	$144,000
18000 @ 16 min × $0.24	$1,013,760
21000 @ 24 min × $0.192	$1,161,216
Estimated revenue	$47,108,576
Less 20% of revenue at wholesale, discounted 25% access or startup, 15% usage	$1,643,809
Total revenue	$45,464,767

The total cost of the marketing and sales efforts can be quantified from the documentation above and recent experience with these expenses. See Table 4.2 for a sample marketing plan estimate of sales lead and commission costs.

If substantial revenues are generated via wholesale business through uncontracted resellers, such revenue is more difficult to commit and forecast, other than through extrapolating trends of existing business. The uncertainty of such revenue is a risk. This can be minimized by documenting the continued marketing and account support programs for resellers in support of this portion of the forecast.

Sales and marketing resources must be separated into retail and wholesale; ideally, a separate *pro forma* income statement is prepared for wholesale and retail operations, or for each individual channel. This shows the higher retail revenues obtained at higher sales costs than the wholesale channel. The

TABLE 4.2   Excerpt from Marketing Plan

Sources of New Customers

Agent A quota at 9 salespeople with 35 sales per month	3,780
Agent B quota at 14 salespeople with 35 sales per month	5,880
New Agent C quota at 6 salespeople with 35 sales per month	2,520
Major Accts. quota at 6 salespeople with 20 sales per month	1,440
Retailer A quota	4,000
Retailer B quota	3,000
Dealers at 10% over previous year	2,188
Resellers at current activity level	4,200
Total gross customer gain	27,008
Leads Required	
50% of agent on direct quotas	13,440
Cost at \$18.00 per lead	\$241,920
Commissions	
Agents at \$350[a]	\$4,263,000
Major accounts at \$150	216,000
Dealers at \$200[a]	437,600
Retailers at \$100[a]	700,000
New residuals at 5% of retail startup/access/usage	\$1,884,343

[a]A weighted average commission based on the channel's mix of customer segments opting for different rate plans that pay different commissions.

return on sales for each portion of the sales channel mix will demonstrate the need to emphasize the most profitable ones (see "Cellular Sales Productivity," Chapter 11).

The development of a credible plan based on projected revenues and costs as they are derived from detailed plans will provide a road map for the year's marketing operations and will be invaluable in the planning process. The more detail that can be developed and the more the participants involved in the plan, the more valuable it will be to the entire team.

5

PRICING

Pricing is one of the basic "four P's" of marketing yet is given very little attention. In addition to its importance in defining margins, it is strategically important in what it communicates to the customer relative to the other elements of the marketing mix.

Price, unlike some popular conceptions of consumer "rip-offs," is not something involuntarily exacted from an intimidated buyer but an amount freely paid by consumers when their perception of value and benefit gained is much higher than the price stated.

Pricing communicates value in its simplest sense when it is lower than that of a competitor for a product or service of comparable worth. Yet it also communicates value when a distinctive product is priced at a premium. A product designed, for example, for high quality and distinctive features should be associated with a premium price. Such a product should not use low pricing as a market entry or market share gain strategy. The advertising and promotional strategy should do that job in such a case, not low price; the distribution strategy should emphasize personal service and knowledgeable salespeople. An alternative strategy, for example, to price below the competitor, would have different complementary strategies in the other areas of marketing.

The determination of price levels, or pricing, as discussed here, follows directly from customer expectations of the value placed on the service uncovered through experience, competitive trends, and market research. Actual pricing levels and structure must fit with the strategic and marketing plans as well as with any customer expectations. Willingness to pay (WTP) is the overall amount potential customers are willing to pay as a portion of the value they place on a service and is divided into structural elements and quantities purchased (access, usage, features, etc.) according to customer need and other inputs to the pricing process. It is assumed that all of the pricing work originates from the strategic plan and marketing plan objectives and strategies. Once the appropriate market research and planning have been done, the basic pricing strategy and levels should almost fall into place by themselves, side-by-side with complementary product, promotion, and distribution strategies.

The *last* step in market-based pricing is to determine if prices determined in such a manner fit within financial, billing and regulatory constraints. It is unfortunate that good pricing strategies are tainted by having been subjected to such constraints before considering customer need. It has been my experience on more than one occasion to have systems people tell me, "We can't bill that way." The mere expression of such a thought is abhorrent in a world supposedly guided by customer need, but moreso by the insolence to believe that the possible is proscribed by the existing.

It would be naive to think that the constraints on pricing, and marketing in general, placed by financial, billing, technical, and other areas are not formidable. They may severely influence pricing in its final form and thereby affect forecast and growth as well. The key point is that these constraints should not be prior restraints on the marketing process. Any substantial changes in pricing levels and structure derived from these constraints require a complete recycling of strategic and marketing plans to revise the strategic position of price among other marketing elements as modified by constraints.

Pricing should have a complementary relationship with other marketing elements and yet of course has many strategic and tactical aspects of its own. Pricing also includes all of the terms and conditions of sale: not only the cost of the service, but the way it is paid, the way it is billed, and the terms of delinquency and collection.

In wireless communications, the customer focus on the telephone instrument rather than the intangible subscriber service has had interesting pricing implications. Customers have noted the rapid decline in wireless telephone prices over just a few years. For customers who do not have a high business productivity need for the service, there is an expectation that the price of the service will continue to come down, merely because the price of handsets has declined most dramatically. The price of airtime has decreased, along with increases in promotional free airtime. But the tendencies of the price of airtime and other services are more diffuse than the highly visible decline of handset prices to zero in many cases. Furthermore, once the price of basic handsets becomes zero, the price of more expensive handsets visibly declines, and free goods and rebates cause the handset price to be effectively less than $0!

Actually, it was the rapid decline of telephone prices that allowed service prices to stay high so long before additional PCS and other carrier competitors entered the market. Customers had been able to sidestep the major price hurdle of the telephone and control service charges by limiting access. Service was an afterthought to the equipment decision, and customers were not as sensitive to service price as they were to getting a good buy on equipment.

These observations are changing as more price-sensitive business and consumer prospects enter the market and as carrier competitors force the airtime price down with price cuts and free minutes. But newer customers are still getting "sticker shock" when they see their first bill, canceling service in many cases.

The major purpose of designing service pricing structures is to use the balance of access and usage to provide value to target customers. The consumer segment wants low fixed charges, while corporations want plans that provide discounts representing the value of the business they bring to the carrier. By combining low fixed charges with higher peak usage rates for "economy" plans and providing discounts to market segments associated with high usage that are tied to customer loyalty, the price structure provides the flexibility that different segments require while lessening the pressure on overall pricing levels.

PRICING STRUCTURE AND LEVELS

The actual development of pricing structures and levels is a key strategic function and must use the information developed in the strategic and marketing plans, combined with other elements required of a pricing plan by customers:

Consistency with customer needs and benefits
Willingness-to-pay (WTP) levels from market research
Simplicity
Certainty
Quantity
Competitive comparison

Pricing must provide the service levels a customer needs to achieve value at prices that meet the constraints of a customer's monthly budget. Prime target market segments for wireless service in the very largest markets might need to achieve, for example, about 200 minutes of usage per month, concentrated

TABLE 5.1   Wireless System Pricing Model

A. Total time usage per month	200 min
B. Peak usage	80%
C. Access as a percentage of total bill	30%
D. Customer willingness to pay	$80
E. Off-peak price as a percentage of peak	60%
F. Access price of total bill	$24
G. Peak usage price	$0.30
H. Off-peak usage price	$0.18

Calculations

$F = C \times D$

$G = (D - F)/\{[B + (1 - B) \times E] \times A\}$

$H = E \times G$

at 80% during peak usage hours, at about $80. Assuming that the user places a value on being reachable and on being able to place calls from anywhere, the access portion can be valued at about 30% of the $80 WTP. Assuming off-peak prices are about 60% of peak, access and usage prices can be directly determined from this information, as shown in Table 5.1.

Assuming that you have similar information for any market, the same calculations can be used to determine preliminary basic pricing. This plan provides the price level and structure desired but does not have any promotional or specific market segment appeal to the target market. These plans and calculations must be modified to take account of free minutes included in the access charge, promotional free minutes, etc., added to appeal to the market segment at which it is directed. The plan also must be verified against financial goals and forecasts to make sure it achieves all objectives. In addition to these tests, the other elements of the pricing plan tests listed previously must be met: Is it simple? Can the customer estimate and calculate the total cost with reasonable certainty? Does it provide incentives for quantity purchase and usage? And does the plan invite favorable competitive comparison?

Additional pricing plans must also be created for other target markets. There are four or five basic pricing plans, each of which should appeal to general market target segments, and named to convey to the prospective user that the plan is designed to be beneficial for his or her type of usage. The five general plans usually offered are variants of the following:

Basic plan
Executive (heavy-user) plan
Corporate plan

Economy plan
Wholesale

The basic plan is actually a nonspecific benchmark plan and may not actually be offered. It is used to incorporate all the structural and price-level assumptions before promotional and segment-specific elements are added as a check on the analysis of other plans.

The executive plan is targeted toward the important small-business executive segment. It packages some amount of usage (usually about 100 minutes) with access, giving a discount of, perhaps, approximately 7% on the packaged minutes. It is important to discount the packaged minutes rather than the excess usage, so that the subscriber perceives a fixed discount. Large amounts of incremental usage over and above the packaged amount are not necessarily discounted (unless desired).

The corporate plan provides an incentive for 15 to 20 or more users to be billed on one bill to one address at a corporation and usually includes a larger amount of usage packaged with access than the executive plan (perhaps 200 minutes). There is a larger discount for this plan than the executive plan, because it encourages multiple users and reduces administrative costs. It may have an associated minimum term of one year and may also include vertical services (e.g., custom calling features and/or voice messaging).

The economy plan is specifically designed for consumers, price-sensitive customers, and those who are not sure about their commitment to wireless services. Rather than a simple discount, its balance of access price to usage price is altered to change the relationship of fixed (access) costs of service to variable (usage) costs for the customer. The economy plan charges an access price approximately one-half of the normal rate and a peak usage price approximately double the normal rate. Thus, the user has the ability to keep charges low by minimizing use of the service and also can control charges by calling off peak as much as possible if he or she has flexibility, as many consumers do. However, the economy plan is named specifically to attract price-sensitive customers. It does not *save* them money at all levels of usage but *allows* them to save money if they manage usage.

There are many variations on these plans, but they have several elements in common. In addition to targeting specific segments, some include discounts and free benefits that recognize the quantity of business that the customer contributes; others provide these discounts and benefits in return for a minimum length subscription commitment, usually a year.

The names used are not idle attempts to merely distinguish one plan from the other. The names should be carefully chosen to suggest the type of user it is most suited for and/or the benefits of the plan. This is important to help the consumer come to a decision, to assure them they are on the plan that provides the most value, and to help position the service as the salesperson presents it to the sales prospect. More sophisticated segmentation might target the security user (light occasional use for emergency and other important calls) and the "wireless road warrior" (heavy user with home rate roaming) with appropriate plans.

Each plan may have an alternative name and a slightly different structure, but there should be a distinctive plan incorporating these themes to appeal to major market segments. It is important not to have too many plans. They should have names that connote to the user the value they receive, or the positioning with a certain style or status (nobody wants to be known as a "heavy user").

It is counterproductive to have too many plans and plans with confusing or nonmeaningful names (e.g., plans, A, B, C, ...), or to switch customers from one plan to another every month by calculating their bill according to each plan and switching them to the least expensive. Such strategies confuse the customer, complicate billing, leave money on the table, and most importantly negate the reason for multiple plans—to enable the customer to easily make a good choice. Of course, the carrier should occasionally check with the user to make sure they are on the right plan to avoid customer dissatisfaction.

It is sometimes tempting to lay out price structures that may attract customers by deceptive packaging; however, in any subscriber service, the customer will soon learn from the bill he or she receives every month if it is different from expectations or if the wrong usage plan is listed. Any short-term revenue advantage of a deceptive plan will be greatly outweighed by long-term customer dissatisfaction. In addition, many marketers have learned that consumers are incredibly intelligent at deciphering pricing strategies and getting the better of the vendor.

Giving away voice messaging free with wireless service, for example, in the hopes of increasing airtime use through retrieval of messages using wireless service will often result in customers using wireless voice messaging in lieu of voice messaging from the telephone company. They will use it to collect unanswered calls from all their telephones, accessing it free from landline telephones to retrieve messages. This is far removed from the carrier's intentions. In many cases, however, some additional wireless use will result from the offer, and the customers' loyalty is increased because they feel they are craftily using the carrier's services to their own advantage.

Thus, it is important in pricing to understand that the wireless carrier should never expect customers to behave in the way that the carrier would desire regarding pricing decisions. On the contrary, they should always expect customers to behave in the way that is most self-serving to the customer and detrimental to carrier profitability.

TABLE 5.2   Illustrative Wireless Rate Plans

Basic plan	Access, $36.00; usage: $0.46 peak, $0.27 off peak
Executive plan	100 minutes included, 7% discount, one-year contract, excess usage at basic rates
Corporate plan	200 minutes included, 10% discount, one-year contract, excess usage at basic rates
Economy plan	Access, $20.00; usage: $0.72 peak, $0.27 off peak
Wholesale	Basic plan discounted 25% on access, 15% on usage (100 lines, 10,000 minutes per month minimum)

Table 5.2 displays each of our illustrative plans against the basic plan derived from our pricing assumptions. Once again, these are sample plans for a hypothetical market, used here to show their derivation and their relationship to each other.

The plans should be designed rationally so that at each level of use, the appropriate plan is the best value for the customer (Table 5.3). Below 60 minutes of use, the economy plan should be the best value; at 100 minutes, the executive plan; at 200 minutes, the corporate plan. The basic plan is priced to be a better value than the economy plan at any level of usage higher than about 60 minutes, but will always be higher than other plans that require minimum usage or a contract length. If plans are not designed carefully, users willing to pay high amounts for usage will migrate to other plans that have been priced lower to attract different types of users.

Because some carriers do not see the value of resellers, they tend to structure wholesale prices (if they have them at all) as small discounts from each element of each retail pricing package. But the reseller plays an important role in adding diversity to the marketplace by providing its own set of retail prices. Wholesale prices should be structured after the basic plan only, but priced lower than any retail plan based on a much larger minimum quantity, for example, 100 access lines, and lower sales and support costs. This permits the reseller the flexibility to lay out retail pricing structures

TABLE 5.3   Comparison of Illustrative Rate Plans

	Included Minutes	Monthly Access	Usage		Minutes of use			
Plan			Peak	Off Peak	50	100	200	300
Basic	0	$36.00	$0.46	$0.27	$57.10	$78.20	$120.40	$162.60
Executive	100	76.26	0.46	0.27	76.26	76.26	118.46	160.66
Corporate	200	108.36	0.46	0.27	108.36	108.36	108.36	150.56
Economy	0	20.00	0.72	0.27	51.50	83.00	146.00	209.00
Wholesale	0	27.00	0.39	0.23	44.94	62.87	98.74	134.61

different from the carrier. Wholesale pricing derived as direct discounts from each retail plan invite the reseller to use copycat plans, and indeed, give him or her almost no room to do anything else. This limits the diversity of attractive features that different providers can use to attract different kinds of customers in the marketplace.

Also, it should be understood that some large companies as end users may also qualify for such rates if they are willing to accept the lower customer service and billing constraints of the plan. The wholesale plan requires certain restrictive features that allow the carrier to price it so low. The reseller (or large multiuser account) pays for access in advance and is billed for a minimum amount of usage in advance; receives only one bill, via tape or disk; has a very short payment window; can have its customers assumed by the carrier for nonpayment; has limited customer service, not accessible by end users; and is subject to other restrictions.

With these restrictions, the carriers' sales, float, billing, and customer service costs are kept very low, and the wholesale business can be much more profitable than its retail business, while affording a great opportunity for independent businesses in the community and diversity of service choices for the customer.

Returning to Table 5.3, the exact number of minutes at which two plans are equivalent, a "crossover point," can be calculated and adjusted based on minutes of use of the segments attracted to each plan. The average number of minutes per user attracted to a particular plan should lie midway between the crossover point with the plan for lower-usage and higher-usage customers.

More or different plans can be created; marketing intelligence should be able to identify groups of potential customers with different needs who might be attracted to different levels or plans. In all cases, the plan intended for a target group should be the most economical at the average or modal level of usage for the group, and only at that level. The basic plan is undifferentiated, and charges under it will usually fall between the economy and the volume discount plans. The wholesale plan should be lower than any retail plan at any usage level.

When we look at sales channels, we will discover another important aspect of the rationalization of pricing plans across market segments. Carriers are discovering, as markets start to mature and attract "marginal" customers— those that use the system less because their need is less, like consumers—that sales commissions and incentives need to be based on the revenue and longevity potential of the customer. Since individual customers are difficult to classify, it is useful to allow customers to self-identify their segment, and thus their usage potential and longevity, based on the rate plan they choose, and for the carrier or reseller to base sales incentives and commissions on the rate plan thus chosen.

This is enormously useful in order to match sales costs to revenues, to compensate salespeople and channels appropriately, and to ensure that salespeople are indeed signing customers who belong to the market segments to which they are directed to sell. However, it places an even larger burden on the pricing plans to be well structured, attractive to the right users, and positioned correctly relative to each other. If misused, and salespeople are, for example, overcompensated for attracting customers to the executive plan and uncompensated for the consumer plan, many light users will find salespeople pressuring them to choose an inappropriate plan intended for heavier users. See Table 5.4 for pricing plans for a typical wireless carrier.

Pricing Plan Variations

In recent years there have been several additional types of service plans available that have become generally accepted as distinctive new structures.

Prepaid. Prepaid wireless service is sometimes considered a product or billing option as well as a pricing plan. Originally intended for customers who could not satisfactorily pass a credit check, it provides service without a bill. It arose as a modification of prepaid phone cards for long distance. The customer pays for service in advance and is told how much time is available each time the phone is used. The airtime balance can be replenished by phone or at a sales point. Airtime is usually priced at a premium to regular rates, especially since there is no monthly access charge. Both network-based and phone-based services are available. In the sense that it requires a separate pricing plan, it can be considered a pricing strategy. Since it requires either system modifications or special phones, it is often considered a new product or service offering. The finance people consider it to be a billing option.

Prepaid has become an important option beyond the credit-challenged. Rather than offer prepaid only to those who have credit problems, it has found its own niche as an important feature for many users. Those who need to control their business or personal budgets to a fixed amount and those responsible for the wireless bills of children or other family members are two examples of applications for prepaid service that have greatly expanded its original purpose and audience.

Home Zone. This type of pricing plan offers both low fixed monthly access as well as low airtime rates within a sector of the carrier coverage area. This plan attracts those customers who are price-sensitive to airtime rates as well as monthly access. Airtime rates are very high outside the home zone, and other restrictions such as roaming denial are often added to make sure the plan is only useful to targeted segments.

TABLE 5.4   Pricing Plans, Feature Options, and Terms for a Typical Wireless Carrier

Service Plans	Monthly Access	Airtime charge	Terms
FirstLine	$25.00	25 minutes free 26+ minutes $0.25	Annual
LifeLine	$15.00	5 minutes free 6+ minutes $0.75	Annual
BasicLine	$32.00	$0.34 per minute	Monthly
Basic TalkLine**	$55.00	100 minutes free 101+ minutes free	Annual
Plus TalkLine**	$79.00	200 minutes free 201+ minutes $0.32	Annual
Ultra TalkLine***	$159.00	500 minutes free 501+ minutes $0.25	Annual

*Add up to two more cellular phones. Charge in per additional line.
**Includes Call Waiting, Call Forwarding, and Conference Calling.
***Includes above features as Mobile Mailbox.

Service Plan	Monthly Access	Airtime Minutes Included	Airtime Charge	Terms
Family Line	$15.00/line			
Any Time Minutes	$25.00	100	$0.25	Monthly
	$45.00	200	$0.22	Monthly
	$99.00	500	$0.20	Monthly
	$350.00	2000	$0.175	Monthly

Corporate Lines	Monthly Access	Airtime Charge	Terms
5–10 Lines	$100.00	0–200 minutes @ $0.35 201–350 minutes @ $0.33 351–450 minutes @ $0.31 451+ minutes @ $0.29	Monthly

Each Additional Line	Activation Fee	Change or Restore	Detailed Billing
$10.00	$30.00	$15.00	$5.00

Custom Calling Features	Charges
Call Waiting	$1.75
Call Forwarding	$1.75
Conference Calling	$1.75
Call Restriction	1.75
One Number Dialing	$4.95
Mobile Mailbox with Pager Connection	$5.00
Mobile Mailbox with FaxMail Connection	$7.50
Add-A-Line	$4.95
All custom call features except Add-A-Line and Mobile Mailbox	$10.00

Source: MobileTel

Included Long-distance and Roaming Plans. Toll charges and premium roaming rates are an irritant to important heavy users who are highly mobile. Premium plans are now including long-distance charges as well as roaming to other cities at home market airtime rates. These plans are meeting with success for a portion of the market representing the carrier's most important customers. The simplicity of such plans is attractive. It is also very useful in providing innovative differentiation of wireless services from landline telephone services as a characteristic of mobility. The continuing decline in long-distance prices and the attraction of its simplicity may make this plan an important factor in attracting new markets to wireless in the future.

Flat Rate. These plans offer unlimited, or virtually unlimited airtime, at a high fixed monthly access price. Such plans are attractive to heavy users because basic access and usage can be budgeted and affordable at high levels. Many of these plans do not include local access, taxes, toll, and other hidden charges that can still make bills much larger than their nominal prices would indicate. Yet these plans are driving down prices for some of the best wireless customers as they are courted by competing carriers with such tempting pricing.

Vertical Features and Options

Vertical features (call forwarding, three-way calling, etc.) have very low penetration in wireless telecommunications, unless given away free of charge. Most calling is outgoing; very few users have need for these services. Salespeople view selling such services as a distraction from the sales presentation and a possible deal-killer if they stay too long after the close to bother with such details. These features are important to customer satisfaction, ease of use, and usage stimulation, however, when presented to the right customers. On the other hand, it is usually not cost effective to try to sell such features to existing customers across the board, but only to help customers who appear to need them when they are in consultation with customer service representatives. Most are priced at about per $1.50 per month; usually they are given away with premium pricing plans. The penetration of voice messaging averages 9% to 15%, and it can be sold at $5 to $10 per month, if incentives are provided, which also increases usage.

Roaming

The use of wireless phones in cities other than the subscriber's home city has been an enormous revenue boon for wireless carriers, especially in larger markets and in those smaller markets immediately adjacent to large cities.

Larger markets may realize 9% to 10% of their revenue through incoming roamers, most of which goes directly to the bottom line, unlike the high cost of obtaining new subscribers. Roamer revenue is growing rapidly with a high sales rate of portable phones to enable business travelers to make wireless service part their air travel plans, as well as improvements in technology, ease-of-use, billing, and fraud control. Markets adjacent to large cities may hardly bother to gain subscribers but merely bill roamer usage of metropolitan users. While roamer fraud is the biggest issue concerning roaming, roaming has several important pricing considerations.

Since wireless calling is mostly outgoing, many users can get along without easily being accessible in a foreign city, although technology is allowing incoming calls to roamers to be passed to them. Roaming is growing rapidly, and soon customers will routinely expect to be reachable in any city. Some crafty wireless customers, however, subscribe to wireless services in areas where monthly access is the lowest, giving them the right to use service in major cities, where access rates are higher. This factor, plus the premium that users place on the value of roaming, should put the nominal roaming usage price at about double that of premium airtime at basic rates. With the advent of higher minimum monthly fixed charges among markets due to included minutes, this differential is no longer as much of a concern. Indeed, many carriers now offer roaming at home market rates for which the average bill from one market to another does not vary too widely. Another way to minimize this differential is to add a fixed monthly price for home-rate roaming privileges.

In addition, because of the cost of clearing roamer billing to the home carrier and the fact that many callers only place one or two calls per visit to a foreign city, many carriers rightly place a fixed daily access or registration charge on roaming. This only becomes a headache to users if they drive through six cities and make as many phone calls in one day or regularly commute between markets; however, it also encourages roamer fraud. With increasing automation of roamer billing and fraud control, such registration charges should become minimums rather than fixed charges and may eventually be entirely eliminated.

Roaming is an artifice created by regulatory barriers. The best situation would be to eliminate roaming altogether, but this might trigger price fixing among carriers. Home-rate roaming and national uniform pricing on the part of some carriers are the first steps toward eliminating special roaming rates.

More importantly, customers who roam are treated poorly by their home carriers as well as by the visited market. Even the term *roaming* is unfortunate as a label for customers. Roamers provide the highest-profit revenue a carrier can make; yet roamers are treated as potential criminals in light of roaming fraud losses suffered by carriers, are given second-class consideration by visited carriers compared to home-market customers directly under their control, and are provided no support in most cases by their home carrier when roaming in another market.

FINANCIAL CONSIDERATIONS

We have been considering the ways in which pricing needs to be structured in order to be consistent with market needs, attractive to particular kinds of customers, and complementary to the other elements of marketing. Our example of the pricing model used particular prices that are only pertinent to a single, hypothetical wireless market and cannot be generalized to any or all real markets. In theory, every market deserves its own pricing structure and levels. Pricing at the same level and structure in multiple markets not only implies ignorance of market need and pricing strategy; it leaves money on the table, and it tells us that the carrier values ease of price setting over profit and growth. Yet pricing plans have become so complicated that when multiple plans of multiple carriers in multiple cites are considered, administration becomes virtually impossible for wireless customer companies. Competitive pressures have also helped to level prices. Therefore, more uniform pricing has become the norm for carriers with multiple markets, and nationwide pricing has become a strategic differentiator for such companies as AT&T Wireless.

The major consideration of pricing for financial purposes is the price level of one minute of peak usage. The start-up fee occurs only once and regardless of its level is not a major element of the revenue contribution of customers when considered over the life of their subscription. The monthly access fee contributes significantly to this figure and becomes more important as monthly usage per subscriber in marginal market segments declines and as more usage is bundled with access.

Off-peak pricing does not stimulate users to move from peak to off-peak usage but attracts nonbusiness users and special segments to the service. It also calls attention to the greater value of peak usage at higher prices and is valuable as a promotional device ("free off-peak minutes"). Off-peak usage provides lower revenue both because of its lower price level and the lower traffic that occurs off-peak.

The peak-usage minute price therefore determines, more than any other element of price, the total revenue as well as the marginal profit of the carrier. The marginal profit of a one-cent increase in peak-usage price is almost equal to the entire large revenue gain achieved. Two negative aspects to raising the peak-usage price are the loss of price-sensitive customers and its competitive and regulatory implications. A simple financial model will illustrate the sensitivity of revenue and profit to peak-usage price over other pricing elements.

Within this context, it is surprising how little empirical experimentation has been done for increasing prices in the wireless environment. Most efforts have been aimed at lowering access and usage prices to stimulate customer growth, to steal customers from the competition, and to provide discounts as an incentive to loyalty. With today's increasingly competitive environment, such experiments are precluded.

The core target market for wireless telecommunications was predicted through market research to have little sensitivity to price. This has been verified and confirmed by the lack of usage and subscriber growth through price reduction. The core target market has changed dramatically in the past few years with the entry of additional competition and the desire to attract new market segments. The general statement is that wireless service has had a low price elasticity and probably still does. The problem today is that an attempt at increasing prices may cause losses to lower-priced competition, but does not cause customers to leave wireless communications.

Yet, because prices are inelastic, reduction in prices will not increase customers and usage revenue enough to offset the decline in total revenue. As a corollary, wireless service is so important to many users that it is unfortunate that an experiment to find the top price that core users will pay for peak without a reduction in total revenue has never been tested. See Fig. 5.1 for airtime prices among several carriers in one city.

There have been some experiments in which service price reductions have stimulated growth; however, such growth is only within price-sensitive market segments such as consumers. A general price decrease gives away money to existing heavy users and stimulates growth in the lowest-usage, most price-sensitive marginal segments at a huge cost in overall revenue. Therefore price reductions must be promotional (temporary) in nature and carefully targeted only to affect new users in segments that are price-sensitive.

Indeed, once competitive carriers have settled into a market, each is rightly fearful of raising prices relative to their competitors even 1 cent. Consequences of such an action include the potential loss of customers and reduction in relative subscriber growth rate.

Certainly one sign that a price increase is in order is an increase in customer growth such that the capacity of the system is constrained and service availability and system performance measures indicate usage congestion. Continued rapid growth during a period of congestion means that the service is underpriced for the value received or that sales incentives are too high.

Los Angeles Pricing (per minute)

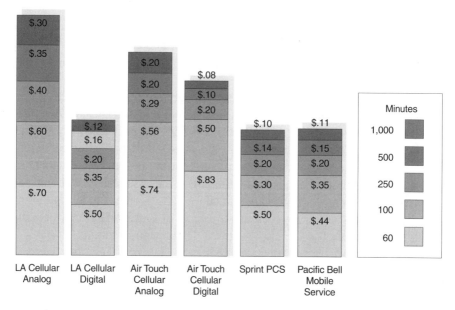

FIGURE 5.1   Airtime prices for a representative city. (Courtesy of Yankee Group.)

Raising prices is an effective way to increase returns and prevent further service deterioration to current customers while system capacity is being addressed. Aggressive growth during such a period is dangerous to long-term customer satisfaction and retention.

While carriers should look at the possibility of the financial benefits of a price increase versus stimulating customer growth, their hesitancy to do so should also point to a need to consider carefully discounts and price reductions as a strategy to increase growth rates, customer loyalty, and reduce churn. A few generalizations might be helpful:

1. Price reductions in usage or access have never been successful in increasing total revenue; with the emergence of price-sensitive segments, price reductions should affect only these potential customers.

2. The basic plan serves as a price umbrella, meaning that it is the benchmark for the levels of retail and wholesale pricing upon which all other plans in the market are based. It can be raised but it must never be lowered, or it will trigger proportionate reductions in all plans in the market offered by all competitors.

3. Any price reduction deemed to be required to attract a new market segment should be carefully limited to a price plan that affects *only* new users and *only* the segment to be targeted. General price reductions merely give away money to new and existing users indiscriminately.
4. Any price change should be tested on a promotional basis in a single market before being permanently implemented in multiple markets.

The recommendation of these strategies should not imply that any price reduction is ever recommended. The continued focus of new customers on telephones rather than wireless service, combined with the monthly introduction of new wireless telephones with improved capability at reduced prices, favors promotional and channel strategies completely divorced from service pricing strategies for attracting new customers. So-called "free minutes" promotions are usually only helpful in trading existing customers temporarily between providers at high advertising and commission costs.

There is no substitute for the continuous evaluation of various scenarios of system growth, market growth, subscriber growth, price changes, and cost changes through a financial model as part of an ongoing overall strategic planning process. Together with a good understanding of historical performance of marketing strategies and other operating variables and programs, the process should demonstrate how price changes differ from market growth and cost reduction in improving financial performance. This evaluation will help providers discover the kind of stimulation in customer growth required to justify a price reduction, especially when the other marketing costs of attracting marginal customers are considered.

Except for short-term adjustments to obtain the full potential of peak-usage pricing, there is probably no opportunity in the long-term future to raise wireless prices further for voice applications. Revenue increases will come from additional services and more efficient use of spectrum; nominal prices for basic voice communication will be pressured downward by the need to fill new capacity and by competition from future wireless services.

From a financial standpoint, payback periods and cumulative discounted cash flows are severely penalized if system capacity anticipates, and capital investment in system expansion substantially precedes, new customer growth. However, competition and customer satisfaction require carriers to avoid system congestion at all costs, and coverage area is a prime consideration of subscribers.

Finally, financial modeling of pricing should assume that any generation of system technology will be replaced every five years; five years ago the assumption was probably seven to ten years or more.

6

SALES CHANNELS

Distribution is one of the four basic elements of marketing. Delivery, or physical distribution, of an intangible subscriber service such as wireless applies only to handsets and accessories, and we will concentrate on sales channels as the most important aspect of distribution for wireless service.

AGENTS

As we mentioned in the Introduction (Chap. 1), the authorized sales agent first emerged as the primary sales channel for wireless telecommunications because of the need for technical radio and installation expertise from the existing radio industry as well as the need for "one-stop shopping" provision of wireless telephones, installation, and service subscription. To begin selling service rapidly, the use of existing sales channels for mobile telecommunications services enabled rapid ramp-up of the sales capability as commercial service began. In addition, the importance of using local talent to participate in the market and the flexibility of paying only for sales performance as a substitute for the bureaucracy and high cost of internal channels cannot be overstated.

The definition of a wireless agent, as used here, is an independent business under exclusive contract to a wireless service provider (reseller or carrier) to sell wireless service at retail to end users in the name of the provider in return for a sales commission. The term *exclusive* refers to the contractual obligation of the agent to sell only for the provider and no other in any one market.

An agent is selected by the provider and is under firm contractual control of the provider in regard to location, services available, sales methods, quota, and quality standards. It usually is a business dedicated primarily to wireless sales as a contractual obligation.

DEALERS

While a wireless dealer may be synonymous with an agent in some markets, as used here a dealer is an independent business that sells wireless service at retail in return for a commission. Dealers differ from agents in that dealers have few contractual obligations to the provider, are not necessarily exclusive, and may change provider loyalties frequently. They may be contracted to an agent (see "broker-agent" later) or reseller. Dealers may also have a contract with the carrier, which usually involves less control than the agent contract, limited use of the carrier logo (in advertisements but not signage), and lower commissions. Such dealers are called *direct dealers* as distinguished from dealers who are indirectly connected to the carrier through middlemen.

Dealers also emerge in cases in which the method of doing business may not match the carrier's or reseller's mainstream sales methods and agent contract requirements. Automobile dealers, for example, cannot meet the strictly controlled terms of an agent contract that may require, for example, a sales floor, an accessory product line, a direct sales force, and/or a technical support capability. While the carrier and automobile dealer may earnestly want to cooperate to provide wireless service, telephones, and installation jointly, the auto dealer cannot be directed about selling methods or where to locate, for example.

Existing businesses whose primary method of doing business is not congruent with a wireless agency can participate in wireless telecommunications through dealer status. Another example would be a retailer, such as a department or consumer electronics store, which cannot change its method of operation to meet the requirements of an agent yet can target the consumer segment much better than the wireless provider's mainstream sales effort. With the emergence of the consumer mass market, the retailer-dealer has become the largest third-party channel for wireless carriers (discussed later). Such channels may have exclusive contracts with providers (*direct dealers*) but cannot operate under the same terms as an agent. Others may be one-person wireless sales entities with no contractual obligations.

BROKER-AGENTS

Broker-agents, sometimes called *master agents* or *aggregators*, have evolved as an additional step in the distribution chain. In addition to their contractual obligations as agents, broker-agents may serve as a clearinghouse for independent dealers. The broker-agents receive wireless subscriber agreements from independent dealers and/or their own contracted dealers, and supply them to the provider along with sales resulting from their own efforts. They service dealers with wireless telephones, installation facilities, and customer service support and pay them a portion of their commissions from the provider.

This relationship provides the independent dealer with faster, local service in large markets, more rapid payment of commissions, and other support that the provider may be slow or hesitant to give to smaller dealers. Since commissions are paid to agents by the carrier as a combination of sales commissions plus residuals on end-user revenue, the broker-agent may pay the dealer most or all of the sales commission (or even a premium over it), in order to retain the residual commission for a more rapidly growing customer count.

The provider gains customers faster in this relationship through an arms-length arrangement with dealers who are too small to deal with directly or may not have the quality standards or desire for allegiance of the contracted agent. Large carriers often have a thick, self-serving contract and a long contractual process that is onerous and frustrating for small dealers to negotiate. Thus becoming a dealer for a broker-agent has many advantages for the dealer as well as the carrier. However, the carrier in this arrangement has less control over the quality of the customer, the professionalism of the sales presentation, or the customer service provided to the end user by the dealer.

RESELLERS

Resellers emerged rather accidentally through the requirement of Federal Communications Commission (FCC) for resale. Various wireless carriers have reluctantly allowed the participation of resellers because of their desire to control end users at retail prices. However, in many markets, resellers are seen as an important part of the distribution channel mix. They may attract certain market segments because of their customer base in other products; they may have a nationally recognized name and an existing customer base in other lines of business; and they may add diversity in sales approach, pricing, or other elements of the marketing mix that the carrier or other providers do not. Most importantly, they may provide for greater profit to the carrier for cases in which the wholesale discount on the carrier's retail prices is less than the carrier's cost to sell at retail.

Many carriers want all customers at retail. Carriers can be uncomfortable with some resellers being carrier competitors or selling for several carriers at the same time. The carrier may be concerned that the reseller will discount the retail price and erode the overall retail price level, or *price umbrella*, for the entire marketplace. Carriers apply pressure to resellers by refusing to supply any marketing support and reducing margins between retail and wholesale. They can tighten credit terms and can force resellers who do not pay on time out of business.

The arguments for using resellers still apply. Wholesale is often more profitable than retail because sales cost savings can far exceed wholesale price discounts, and the carrier can set discounts based on sales cost savings. There are many more strategic and less direct reasons for encouraging resale. Introducing any types of new channels (profitably) is important for two reasons: first, to reach new segments in ways that existing channels do not, for example, by offering alternative promotions, price plans, or levels of customer service; second, channels must continually expand or the carrier will stop growing due to churn.

With the increasing emphasis on regional coverage and nationwide service as competitive advantages, carriers desire to resell in markets in which they do not operate a network. New handsets permit dual-mode, trimode, or dual-frequency operation with multiple carrier technologies. Many resellers offer nationwide service also. Thus resale is becoming an important strategy for carriers as well as resellers.

RETAILERS

Retail stores may participate in wireless businesses as dealers as described previously or as a special type of agent or reseller. Because of their increasing role in wireless commerce as the consumer market grows, they should be treated as a separate category of sales channel.

Retailers become an important factor in a particular market as a sales channel due to three factors:

1. The rapidly declining effective price of wireless phones, which is the primary barrier to consumer adoption of wireless
2. The increasing consumer knowledge of the benefits and use of wireless through word-of-mouth advertising, without the aid of a direct sales force or heavy advertising effort

3. The need to match the lower average monthly wireless revenue and shorter customer life of the consumer market segment with a lower-cost marketing and sales operation.

Retailers now play a large role in wireless distribution. The emerging consumer market expects to buy wireless phones in consumer electronics stores rather than from traditional agents, dealers, and carrier stores. Retail consumer channels account for about 20% of sales, and this number continues to grow rapidly. Based on the lower requirements for service and sales effort, it would be logical to assume that commissions for retailers would be lower than that for traditional agents.

But while carriers are paying traditional agents lower commission rates for customers who choose consumer-targeted rate plans (see Chap. 5, Pricing), they are paying higher commissions to retailers because of their draw of consumer customers. If the retailer truly is bringing in a lower-quality customer as the consumer, commissions should match the lower revenue potential. However, competition among carriers for the two or three largest consumer electronics retailers in a metropolitan area increases commission rates competitively.

The payment of standard commissions to attract a less desirable segment may be a necessary short-term strategy to capture these agent and dealer channels for long-term planning. However, individual agents or dealers should never be paid rates higher than the established rates for levels of performance for other channel members in the same category. Otherwise would result in the requirement that all members of the channel to be paid the higher rate, without fail.

Retailers have the highest churn of any channel. They have the most consumers—the lowest-need, shortest-lived, lowest-usage segment in the wireless business. In addition, they usually do not want or have residuals in their contract, so they have no incentive to keep customers.

National retailers like Sears, Circuit City, and Radio Shack will insist on working directly with the carrier merely to assert their importance, even if other channels offer good terms. National retailers have the clout to get even better prices than the carrier on phones. Local and regional retailers, even if large, will use whatever provider pays the highest commission (usually without residual) and provides low-priced phones, demonstration lines with free usage in stores, and cooperative advertising money. In either case, standard agent contracts do not work.

The carrier will normally provide full training to retailers. The problem is that the retailers do not want it. They like to obtain information from carriers at national headquarters and distribute a memo to all their stores. If the carrier offers local training, either no one will attend or those trained will soon be moved to another department or store.

INTERNAL DIRECT SALES FORCE AND MAJOR ACCOUNTS

The carrier organizes its own general direct sales force separately from its major accounts direct sales force. The direct sales force normally does not specialize based on industry, but may form several geographic branch offices with installation facilities or use agents or outside contractors to do installations. This sales force is formed with the intention to compete in the same market segments as agents and resellers: small business and general business, with a concentration in professional services, construction, real estate, and financial services. These sales forces are created to take business from resellers and agents at a lower cost or higher revenue; thus they create rather than resolve channel conflict.

In addition to the cost issue, the carrier wants to control accounts directly if it suspects that agents and dealers are churning customers among carriers and resellers. Some carriers say they have shown that when the direct sales force controls customers from the same segment, they stay longer and have higher usage than agent-generated customers.

A key account for a carrier is a large corporation or *major account*. Such accounts usually include the local offices of the Fortune 500 companies, local firms of large size (over 1000 employees or $100 million in annual revenue), and government accounts. A *key account* for an agent, on the other hand, is any account that has 10 or more users. Key accounts are an important source of referrals and additional business. Easier administrative procedures, special billing, and usage reports are important incentives to offer major customers, but special pricing plans are most important.

The major accounts group is usually broken down by industry within the companies that fit the definition just given. Each representative specializes in an industry or a few large companies and gives multiple presentations to demonstrate applications pertinent to the company's industry that increase productivity through wireless telecommunications. Large accounts often require three to five presentations before they enter into an agreement. However, once the account is established, the repeat business is sometimes so good that the carrier has to take the account away from the salesperson because he or she makes too much money with little sales effort, ignoring the other accounts in his or her industry specialty.

Large companies were slow to implement wireless telecommunications on a company-wide basis but by 1987 began full-scale adoption of cellular services. While the carrier offered easy administrative procedures for subscription and installation, it also offered phones at discount prices but above wholesale cost. Because of the rapid decline in phone prices, soon this was no longer an important advantage.

Today, price discounts on service are important. In California, deviations from published tariffs are only permitted for government agencies, for which the price is sometimes discounted 50% from retail.

Commissions to agents are an explicit sales cost, and when commissions increase the carrier may consider using its direct sales force in competition with its agents. Considering the carrier's overhead, it is almost always better to use external channels instead of an internal sales force. The exceptions are major accounts and direct marketing to consumers.

DIRECT MARKETING

Direct marketing is the ability to generate a sale directly as a result of a telephone call or mail solicitation without a sales call or store visit. As awareness of wireless telecommunications in the mass market increases, this channel becomes more and more viable. Because it requires a very carefully constructed promotion and direct mail or telemarketing team, it is usually only within the power of the carrier to develop a viable program. This channel has great benefits for the carrier by providing reduced sales costs and commissions, direct control of the sales process and the customer, and increased satisfaction of the customer through rapid fulfilment of the sales transaction.

The carrier may merely offer a low-priced portable telephone or a free premium with the telephone and service subscription. Response is almost entirely by telephone, although printed advertisements usually offer a reply card also. The carrier may arrange to deliver a programmed portable telephone by overnight express with initial or permanent billing through a validated credit card transaction.

Direct marketing includes telemarketing and direct mail, with which either results in the close of a sale. A distinct set of telemarketing talents is required in this environment. In addition to being a special case of sales channels, direct marketing is also distinguished by its reliance on a feedback mechanism of productivity measurements and a methodical working of a computer database of prospects; for this reason it is often called database marketing today.

This particular channel has increasing importance, because portable phone sales exceed 95% of new wireless telephone sales, and general consumer awareness of wireless services is rapidly increasing. Other special forms of this channel include catalog and infomercial marketing, which address specific niche segments of consumers.

INTERNET MARKETING

Internet sites run by carriers as well as resellers that sell wireless phones and service are rapidly appearing. This channel can be considered a special version of direct marketing, in the sense that no face-to-face contact is used. However, the target is the Internet user who is directed to the site, rather than a selected, high-potential consumer who is contacted by the seller. Such sites are still subject to problems of responding to questions, the desire of customers to select the telephone instrument physically, and the lack of local, timely promotional offers compared to other channels. This channel is emerging because of the proliferation of the Internet rather than consumer demand, and no conclusions regarding its effectiveness or success have yet been offered.

COMPANY STORES

Wireless carriers in the past have created their own full-service sales and installation centers with limited success. Such centers have difficulty in controlling costs to be competitive with independent dealers and retail agents; they cause friction in the retail channels because the carrier is competing for the same customer as the agent.

Unless there is a severe lack of potential agents and dealers in a market, company sales centers usually do more damage than good. They do not capture significant market share and compete directly with other channels. These centers operate at higher expense that the independent channels, the costs of which they are trying to cut.

The most recent efforts of carriers to control channels to the end users directly is the emergence of carrier-owned retail stores (see Fig. 6.1). Such retail stores have the same problems as their full-service center predecessors. These retail locations are feasible because the services required at the point of activation have been minimized with the domination of portable phones and simple programming. These sales outlets may appear to lower carrier distribution costs. However, the management overhead, logistics expenses, and other costs created are not well anticipated by carriers who wish to have their own stores.

The company store concept also entails the problem of establishing a new channel for a retail good directed at consumers. Specialty stores are usually

FIGURE 6.1 The environment of a carrier store. (Courtesy of PrimeCo Personal Communications, L.P.)

geared to customer needs and deal in an entire line of products for a specific purpose: pets, auto parts, or furniture, for example. Consumers tend to group together consumer electronics and telecommunications and expect to find these items at full-line electronics stores not boutiques. Wireless services stores are not "destination" stores and have difficulty generating traffic in standalone locations outside malls. Inside malls, the space is very expensive.

Thus the concept of a retail store for wireless services only does not fit with the expectations of a large portion of the consumer market. It takes a lot of advertising to drive potential customers to the stores for the specific purpose of wireless services. While it is important to have a storefront operation at agent and dealer locations to accommodate the consumer market, carrier-owned retail wireless specialty stores for consumers may not be cost-effective. Such stores are rapidly growing, however, because carrier competition limits the availability of good third-party consumer outlets.

On the positive side, carrier stores enable the carrier to control completely the environment, sales process, and every aspect of the presentation of the product to the consumer market. Thus, stores can be created with a special emphasis on merchandising and presenting wireless products. Carriers report that customers of their retail stores have higher average revenue per user (ARPU) and continue to return. However, this phenomenon may be a result of the demographics of a person attracted to this kind of channel, rather than the result of the quality of sales presentation in the store.

CHANNEL MIX

Figure 6.2 shows the various possible relationships of wireless distribution channels using only general channel classifications. Wireless services can support a two- or three-tier sales channel arrangement because of the high number of small dealers and other small channels requiring local support, and the high commissions paid as a combination of up-front and residual payments provide enough incentives to share commissions through the channel. If the relationship of wholesale to retail prices is adequate, resellers can support their own independent sales channels below the carrier level.

While some channel strategies may be recommended over others, the only true formula for success is to use a mix of channels, each of which concentrates on a different kind of customer. While there will be a broad overlap of segments that do business with each channel, the goal is to have channels that appeal to customers in different ways.

The carrier's direct sales force provides the type of commitment and support that large corporations require. Traditional agents and dealers will find customers in the small-business and consumer segments with the full complement of installation and technical help. Retailers target the mass consumer market. Resellers individually attract diverse markets such as ethnic groups, associations, factory-installed wireless units in new cars, and price-sensitive subgroups in many segments with price structures that differentiate or discount from the carrier.

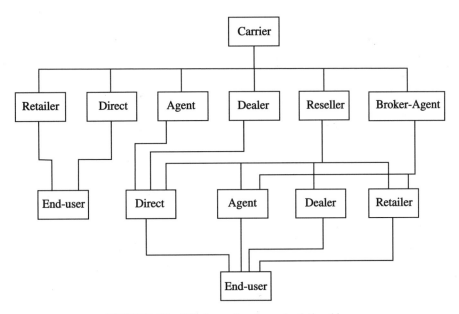

FIGURE 6.2   Wireless sales channel relationships.

MARKET SHARE BY CHANNEL

It is dangerous to attempt to portray a "typical" share of market by channels, but it is important for a carrier to understand the shares locally for all carriers. The following figures assume a market with reseller activity and an internal direct sales channel that handles only major accounts:

Retailers	40%
Agents and dealers	15%
Company stores	20%
Major accounts	10%
Resellers	10%
Telemarketing	5%

SALES CHANNEL CHARACTERISTICS

Sales channels can be classified in order to distinguish their roles in the wireless marketplace. See Table 6.1. Because wireless telecommunications is lucky enough to attract diverse interests as potential channels, it should cultivate all of them for current and future use. Many wireless carriers have been less successful trying to control distribution through direct channels: they have failed in effectiveness because they could not cover the market and failed in efficiency because their costs were too high. Rather than assume that they need to control customers through internal channels only at retail, carriers need to examine the tradeoffs of control versus profit, growth, and other considerations to determine objectively the most appropriate mix of internal and independent channels.

By knowing the channel preferences of each target market segment, the cost per sale of each channel, and the customer volume each channel can deliver, the carrier can devise a hybrid channel strategy that most effectively and efficiently delivers the market. No single channel can do both by itself.

There are other benefits to multiple sales channels, including better coverage of market segments not attractive to retailers, mainstream agents, and direct sales forces; multiple daily advertising presence in the media; additional price competition on equipment as well as service and promotions; and the ability to have enormous resources devoted to locating wireless customers with the external channels paid only for actual sales performance in obtaining new customers.

The agent or exclusive dealer emerged first as the predominant sales channel for wireless services. The ability to sell a small-business executive

TABLE 6.1   Sales Channel Characteristics

Category	Major Accounts	Agent or Dealer	Reseller	Company Store or Retailer
Target Market	Large accounts, government	Small and medium businesses	Small and medium businesses	Consumer
Compensation	Salary and low sales commission	High sales commission and residual; profit on accessories	Margin; profit on phone sales	Sales commission; profit on accessories
Branding	Sells in name of carrier	Sells in name of carrier	Sells in own name	Sells in name of carrier or self
Risk level	Low risk	Medium risk	High risk	High risk
Return	Low return	Medium return	High return	Medium return
Investment	None	Medium	High	High for a company store; None for a retailer

on a single direct sales call, and to promise rapid activation, a quality mobile phone installation, and local, personal service on a flexible basis, worked in favor of this channel. This agent channel has now adjusted to become a storefront operation and serves consumer as well as small business customers.

The carrier's own direct sales channel usually works quite well for major corporate accounts but is inflexible in pricing, terms, sales approach, and rapid competitive response required to work the general business segment in most local markets effectively by itself. Agents and dealers make daily adjustments in phone prices, commissions, and incentives to finalize deals under fast-changing local conditions, a feat impossible to manage in a larger, market-wide organization. The position of the agent rapidly eroded, however, with the penetration of additional segments like large corporate accounts and consumer markets. Corporations require patience, staff work, repeat visits, professional presentations, volume discounts, and account servicing that only the carrier can support through its major accounts sales force.

The consumer market, which expects to travel to a nationally known retailer to purchase a portable phone requiring no technical assistance or installation at a discount price, cannot support, and does not require, the kind of direct sales effort and services offered by the agent. If carriers and resellers are willing to pay the same commissions to a retailer that they are prepared to pay an agent, the retailer can discount the telephone at or below cost, subsidized by the commission (where permitted), because its sales costs are so low and spread over so many other products. Thus, with the rapid growth of the mass consumer market, this channel has equalled or taken over the dominance of the agent and dealer channels.

CHANNEL CONFLICT

Because of relatively high commissions paid for service activation by carriers and retailers to agents and dealers in lieu of nonexistent profits on telephone sales, carriers and resellers may direct their sales forces to compete directly with their agents and dealers. Such practices do not recognize the implicit cost of internal direct sales forces (benefits, salaries, supervision, etc.), which, when added to the lower explicit sales commissions paid, make the total cost of internal sales forces higher than external channels. The internal sales force should be directed at specific market segments that do not interfere with the efforts of agents and dealers directed at the general marketplace and core market segments.

Carriers may also reserve leads generated by promotions or telemarketing for their own internal salespeople. This forces internal channels to be in direct competition with their agents and reduces the effectiveness of agents. If unchecked, agent salespeople will eventually be directed mainly at difficult prospects and cold-calling, while more expensive and highly trained internal sales professionals will be directed at easier sales at much higher costs. Each channel needs to be supported toward success and paid according to performance as directed toward specific market segments or it will wither and die. Sales management should treat both internal and external sales channels alike as partners in success.

SALES COMMISSIONS

Sales commissions for wireless subscriptions are higher than they might be if there were no competition and better profits on wireless telephones, which were intended to be a large profit center for agents and dealers. Competing dealers may help drive commissions downward, but competition among carriers and resellers for the best exclusive dealers and retailers drive them up more than other factors.

Sales commissions to agents are structured based on a cash sales commission at the time a subscriber is placed on the service, and an optional residual commission is calculated as a percentage of access and usage revenue (excluding toll, which usually has no markup). Typical up-front commissions in the top 30 markets may range from a low of about $100 to a high of $400. The mode is around $200, because of the weight of low commissions paid for low-use consumer users and government accounts. Residual commissions may range from 2% to 7% of revenue, depending on the market and the size of the agent's base of customers, and paid for two to five years or the life of the customer.

Many carriers and resellers have resisted paying residual commissions even to their best agents and dealers, because they have difficulty in relating the payments rendered to actual services provided or results in longer retention of customers through such incentives. However, they have more recently recognized some of the important benefits of residual commissions:

1. The residual commission becomes a major component of the dealer's revenue stream after the dealer has sold the first few thousand customers. It becomes the carrier's "insurance" that the dealer or agent will never leave a carrier for a competitor and lose future residual commissions on existing customers.

2. The residual supports the overhead costs of the agent through slow sales periods. As a one-product direct sales operation, agents are extremely vulnerable financially to changes in sales levels. This effect protects the carrier by ensuring a quality agent's ability to provide continuous quality service.

3. The residual commission provides an incentive to the agent to prevent customer loss and to secure high-usage, long-life quality subscribers.

The residual structure has also stimulated the development of the broker-agent in larger markets. As defined previously, the broker-agent passes the sales commission from the carrier through to independent dealers or uses it to provide wireless telephones to small dealers or one-person sales shops at subsidized prices. The broker-agent makes money on the climbing residuals from a rapidly expanding customer base achieved through multiple dealers. This operation becomes a satellite support center for dealers, with much more flexible and local services than the carrier may provide and faster payment of commissions in cash.

The combination of a commission structure that supports several tiers of distribution and subsidization of the price of a phone and a focus of the customer on the telephone decision rather than the carrier decision results in an environment in which the carrier alignment of the channel that sells the phone determines the carrier to which the customer subscribes. Thus the carrier needs to use residual commissions as a tool to gain and keep the loyalty of the agent, retailer, and broker-agent.

RESELLERS

As mentioned, the role of resellers arose with the requirement of wireline carriers to allow nonwireline carriers to sell wireline service until the second nonwireline system was in operation. Thus many of the wireline carriers have viewed resale as a burden that furthered the interests of their future competitor. They have tried to contain resale margins in order to minimize the penetration of their competitor and have provided only minimal support of resellers.

Many carriers have not recognized the benefits of selling at wholesale. It not only relieves the carrier of the sales costs of an internal sales force and commissions to dealers and agents, which alone may be higher than the margin they give to resellers:

1. Resellers pay in advance for access and usage and may relinquish their entire customer base to the carrier if they default on payments. The cost of float, collections, and bad debt are all virtually eliminated for the carrier.
2. Resellers and their sales channels provide all customer service to their end users; the carrier needs only to provide customer service to one contact at the reseller for system activation or deactivation of subscribers.
3. Resellers provide all retail billing; the carrier only needs to supply the reseller a tape or disk of billing transactions on a timely basis to one address.
4. Resellers provide a different set of offerings, prices, promotions, and terms to customers than any carrier. As an alterative provider in their own name resellers make the local provision of wireless service appear more like an industry than a near-monopoly. Resellers such as MCI, GTE, and Motorola bring a national reputation to a local market.
5. Resellers access different market segments in different ways from the single retail approach of the carrier. They may also have a dedicated customer base from an unrelated product line as potential wireless customers that the carriers cannot target directly.
6. Resellers control customer acquisition, billing, collections, and bad debt carefully and thus provide a better-than-average quality of customer because their business is entirely based on financial efficiency.

Because of their desire for retail control of their customers, and, for wireline cellular carriers, the heritage of having their competing carrier as their largest

TABLE 6.2   Characteristics of Agent versus Reseller Candidates

Agent Candidates	Reseller Candidates
Wish to work under the carrier's name	Wish to work under their own name
Willing to work under the carrier or reseller's rules	Want to develop their own rates and business terms
Have low risk tolerance	Have high risk tolerance
Work for commissions	Work for margin or float
Require $0.5 million finances	Require $1.5 million finances

wholesale customer, many large cellular carriers have traditionally discounted the value of resellers without examining the profit potential of encouraging and supporting them. It has been shown through analysis of documents filed with regulators that for carriers in California, retail operations may be losing money or are barely profitable, whereas wholesale operations are the main source of profit for the carrier.[1]

AGENTS VERSUS RESELLERS

Agents are primarily sales organizations that are interested in using the carrier's name, logo, and support to promote sales, making money on retail markup of the wireless telephone and installation as well as commissions from the carrier for new subscribers. Resellers are organizations that want to sell in their own name. They require an efficient billing and collections process and require more capital and financial wherewithal to get started. Thus these characteristics help distinguish independent companies as candidates for different kinds of channels (see Table 6.2).

NATIONAL WIRELESS PROGRAMS

MCI, Motorola, and GTE are major companies that are national resellers. Other, smaller companies also are national resellers in major markets and make it easy for customers to obtain service for any market from one provider.

AT&T Wireless was until the mid-1990s the only major company with a retail national accounts program for major corporations. Such programs offer low-priced telephones and custom billing; rates are determined by local

---

[1]"Car-Phone Rates Run into Static", *San Francisco Chronicle*, May 8 1989, p. C1.

market discounts, with more and more national rate programs available. Now, AT&T is allowing independent carriers to become affiliates. They market their service under the AT&T name where AT&T does not offer service.

AGENT ARRANGEMENTS AND CONTRACTS

Agent Organization

Agents are typically organized as follows:

	Personnel	*Percent*
Sales	20	56
Telemarketing	5	14
Installation and technical	3	8
Customer service	3	8
Management	3	8
Clerical	2	6
	36	100

All salespeople are generalists, and none is confined to certain industries or account sizes. The entire business is devoted to wireless services only. There is one, or perhaps two office locations within a market, and agents usually only operate in one market.

Agent Sales and Marketing Activities

Qualified sales leads are generated 40% from carrier advertising and telemarketing, 20% from internal agent advertising and/or telemarketing, and 40% from referrals of existing customers and the salesperson's own networking. Referrals are most effective with carrier advertising leads second in importance. The internal telemarketing group uses lists from listing companies that are coordinated with the carrier to avoid duplication. The carrier reserves leads to major corporate accounts for its own internal major accounts group.

The carrier is the primary advertiser for wireless service in the marketplace and sometimes supplies sales leads to its agents. It reserves leads to major accounts, government, education, institutions, and associations for itself. General leads are shared with agents and the internal sales force (if there is an internal sales force for other than major accounts) on a pro-rata basis according to the size of the sales force or sales quota of each. The carrier designs promotions in cooperation with its agents based on featured telephones, assuring that the manufacturer can supply them to agents at an agreed price. The carrier supplies sales training that the agent supplements. Most marketing brochures, forms, and "leave-with" fliers are designed by the carrier and supplied to the agent.

Agent advertising is usually confined to small newspaper ads in the business section promoting featured wireless phones. The carrier uses newspaper and radio ads; and TV is used in smaller markets for which it is less expensive. The carrier provides "co-op" (pays a portion) of media advertising based on recent sales levels of the agent but must approve the advertising layout and copy.

A direct salesperson used to average three sales per week, or 12 to 15 per month. With greater customer awareness provided as markets reach maturity and sharply lower telephone prices, a salesperson in a major market now can be asked to average 35 sales per month at $80 commission per sale. A salesperson's only other benefits are wireless telephones and service and a car allowance. The commission may not be constant but may include some portion of the markup obtained in the sale of upscale equipment as well as the activation commission. The sales cost per activation is about $250 and fully loaded is about $400. Salespeople receive no salary but sometimes an advance drawn on commissions. At this sales rate, agent direct salespeople earn about $2,400 to $3,000 per month.

Value-added enhanced services (VAS) are sold by direct salespeople, if at all, only at the time of sale of the telephone because such sales are too expensive otherwise. Such services are not even mentioned unless a need is uncovered during the sales presentation. Salespeople like to keep the presentation simple and leave as soon as possible after a successful close. Paging is generally not sold by agents. Although there is an increasing strategic relationship among the services of wireless voice, paging, and voice messaging, paging has not penetrated wireless voice users more than about 20%. The primary markets of executives, managers, and consumers are not associated with paging; other segments usually already have it when they move up to wireless telephone service.

Customers

As mentioned, agents have a rate of 35 sales per salesperson per month, or about 700 sales for a sales force of 20. In a mature large market, the typical agent's customer base will level off at about 15,000. At this level, at a churn rate as low as 3% per month, the agent will lose as many customers as it gains if it does not expand its sales force, increase its productivity, move into storefront sales to consumers, or act as a broker-agent.

The agent targets the small-business category for 50% of its sales; these customers are most likely to buy on a single sales call from a direct salesperson. Another 30% comes from midsize businesses and business users of large companies not using the corporate account. Consumers comprise most of the remaining 20%. Consumers do not prefer a direct sales channel, are hesitant to buy on a single sales call, and are more price-sensitive than business customers.

Consumer sales are 98% portable phones. Large accounts are usually reserved for the carrier internal sales force; conflicts are negotiated on an individual basis. Consumer sales through agents have leveled off at 20% because consumers are increasingly attracted to retail stores for portable handsets. The revenue per month for a consumer is less than $30 compared to $60 more for a business user. In addition to lower usage of consumers, consumer pricing plans provide a decreased monthly access fee combined with a higher peak-period usage rate. This also accounts for part of the lower monthly average consumer billing.

Pricing and Equipment

Leased equipment and service were previously often bundled in advertisements but broken out separately in billing. Because prices for telephones have decreased greatly, bundling of equipment is no longer a major consideration. Instead, agents discount the phone with part of the sales commission or sell it at cost when they cannot get full price. In California, the state public services commission (California Public Utilities Commission or CPUC) several years ago prohibited the subsidization of the telephone price with commissions by requiring that telephone prices cannot be contingent on service subscription, in response to complaints of resellers who cannot compete with carrier commission levels. Agents and retailers still discount the phone without requiring service subscription, assuming they will get 80% of the activations anyway.

Current wholesale prices for wireless telephones are about $75 for basic sets and $300 to $600 for the latest pocket portable models. The direct sales force of the carrier or agent will try to price the telephone about 5% over wholesale when selling to businesses. For consumers, agents and dealers usually offer telephones below wholesale, subsidized by part or all of their sales commissions, usually through retailers for which the channel sales cost is low.

PCS carriers have introduced the concept of the no-contract dealer along with the no-contract subscriber. The carrier provides the retailer or dealer with a phone-in-a-box to sell for $150 retail. This does not require the retailer to break any exclusive contracts to act as an agent or dealer. The retailer or dealer provides no activation or service, so it cannot be accused of this. It is merely selling a new product. It makes a one-time profit of $50 on the phone, effectively its sales commission. The user subscribes by calling an 800 number on the cell phone (over-the-air activation). The process is advertised to the consumer as a no-contract, no-hassle activation. Like other programs, it is now imitated by non-PCS carriers.

This is a good example of how a creative, innovative program can result from a problem. It starts with the problems PCS carriers have in getting good third-party retail dealers if all of the good ones have been taken. In addition, what unique customer proposition (see Chapter 9, "Unique Customer Proposition") can PCS carriers make to consumers? The solution to the distribution problem is also the solution to the marketing problem.

Originally, carriers engaged in equipment sourcing to obtain discounts that their agents could not get. However, manufacturers soon offered almost the same discounts to agents and dealers at low quantities, and carriers did a poor job of managing telephone inventories. Carriers now often buy equipment only for themselves and merely guarantee quantities for their agents and dealers rather than handle the equipment. Carriers may contract to label an exclusive model with their logo for delivery only to their agents. Manufacturers and wholesale distributors manage inventory and service far better than carriers. Major distributors have evolved into complete logistics services for carriers, handling such details as branded phone and accessories packaging and fulfilment of direct marketing sales orders to consumers (delivery of the phone).

Manufacturers supply agents with product literature, demonstration models based on sales levels, and co-op at about 4% of sales. Larger agents and retailers usually buy directly from manufacturers, and buy from brokers or wholesale distributors only when an odd brand or rapid delivery is required for an out-of-stock unit. Motorola requires dealers to go through distributors.

Dealers (or sub agents) and retailers typically sell below cost in order to make commissions on activation. The direct sales force of agents and carriers tries to recover logistics costs on the phone but will sell below cost when necessary. The broker-agent passes the phone through to subagents (dealers) at cost in order to get an activation from the subagent. At about $300 initial commission plus residual commissions, everybody makes money.

Service

Customers have learned to refer usage and billing questions directly to the carrier. About 300 customer service calls per month, separate from registration and subscription, are received by the average dealer, distributed as follows:

Additional work on telephone activation	10%
Question on new equipment operation	50%
Dropped calls or poor reception	30%
Billing or other	10%

Follow-up service required by customers of dealers and retailers rather than the carrier is minimal and is not proportional to the residual commission. Almost all customer service rendered by the dealer relates to equipment and initial activation and subscription. The carrier resolves billing and other questions that have been determined to be service-related rather than equipment-related. The dealer or store selling the phone handles all equipment-related problems If you ask a carrier or dealer who customers call first for service, each will identify itself as the customer's preferred first point of contact.

Business customers expect a sales call within 72 hours of a request and activation within three business days from purchase. A record is kept for all customers that includes an entire sales and service history. This is used for customer support as well as new customer follow-up and periodic courtesy calls for service and new customer referrals.

Agent Contracts

Agent contracts are standard for the traditional agent, which is a full-service facility dedicated to wireless telecommunications with a direct sales force and retail storefront operation. The only possible variation is the commission level. Even though this might be set out in the contract according to volume, a carrier might pay a higher commission for the same volume to get an attractive agent. This is not recommended, as it eventually forces the carrier to pay that rate to everyone. Of course, the commission rates change during the life of the contract many times anyway due to changing market conditions.

A typical monthly quota for a new agent would be 200 gross new subscribers. This is based on a sales force of six, each making 35 sales per month.

A standard agent and a broker-agent can have the same contract. Sometimes the standard contract will be modified to state how much of an agent's business must be direct versus business through agents in order to assure a high concentration of direct sales for certain agents. Broker-agents may get a lower commission because they may be involved with dealers who churn customers from one carrier to another or otherwise have no control of the quality of the end user.

A contract with a retailer is different, because the carrier cannot control a retailer to do business as the carrier wants. A company such as Sears or Radio Shack will dictate to the carrier its terms, rather than the other way around, and a custom contract usually has to be designed for each retailer, based on a model called an *authorized retailer* or *authorized sales center* contract. Retailers will often not allow any provisions regarding any required services performed by the retailer as agent, any language about the way the retailer may be inspected or supervised by the carrier, any restrictions on which wireless telephone product lines must be sold (or requiring them to carry private-label merchandise of the carrier, among others), nor any quota requirements. The carrier is usually happy to get retailers even on these terms because of the value of their own name and presence in the consumer segment.

Agency Terms

The agency contract is virtually always exclusive with one carrier; otherwise the entity becomes a dealer under a loose contract or no contract at all. The commission terms, based on the sales quota and previous performance, have already been described. The contract usually describes the requirements of the agent in very specific terms and assumes that the agent has been carefully accepted for its role by the carrier or reseller:

- The size and location of the agent's facility, and the operations it must perform in terms of sales, customer service, installation, equipment inventory, computer systems, etc., are stated.
- The agent is required to make wireless service its primary or only business at the location and subscribes customers exclusively for one carrier or reseller. The location may need to be approved by the carrier.
- The agent must have advertising and promotions approved by the carrier, and the contract prescribes the way in which the agent may use the carrier's name and logo.
- The contract defines the sales quota, quality standards, and commission structure for the agent. It also describes the conditions under which failure to achieve quota and other performance deficiencies (customer complaints, sloppy operations, etc.) may lead to severance of the contract.

Commission terms vary based on the quota and customer base of the agent rather than the type of customer. Residual commissions based on customer revenue automatically regulate the quality of the customer. Recently, carriers have been trying to implement programs to lower commissions to all agents for consumer customers versus business customers because consumer usage as a class is so low. In addition, carriers are denying commissions to their internal major account salespeople for sales that are not to major accounts. Those markets paying the high end of commission levels are continuing to lower them because of a decline in usage per customer.

An agent would have to have more than 4,000 to 5,000 customers to get a residual of more than 5%. Commissions are lower below the top 60 markets but not too much lower, because all over the country the low prices of phones do not provide margins to agents to support them. Commissions would be about $300 in a city like San Francisco and $200 in smaller cities beyond the top 60.

The Ideal Agent Contract

An ideal agent contract might start with a high up-front commission to attract new agents and move quickly to a high-residual, low up-front commission structure to keep them once they start to accumulate a customer base. This contract would also give agents incentives to keep existing customers rather than provide incentives for their salespeople and subagents to churn existing ones.

The ideal long-term contract would have commission rates of about $150 to $200 per activation, and residual commissions of 3% to 7% of revenue depending on the size of the customer base. The actual contract might actually start out with up-front commissions of $300 for one year with no residual commission and then move to the long-term model. It is important to require customers to stay on the system for 180 days for commissions to be earned.

Sales of a pricing plan geared toward consumers (low access, high peak-usage rates) generate up-front commissions 50% less than regular. Sales of a pricing plan with a 200-minute minimum per month or a one-year minimum length provide commissions 20% to 25% higher than regular.

The agent contract provisions should include:

1. Agent exclusivity
2. Full-service facility
3. Increasing residual commissions with growth in customer base
4. Agreement to leave major accounts to the carrier

5. Agreement to leave general business accounts to the agent
6. Requirement of some agents to concentrate on direct sales versus subagents
7. Requirement that only the agent direct sales force may use the carrier logo, not subagents of the agent
8. Provision that the carrier may offer wireless phones and other offers to its existing customer base or to the general public on a direct marketing basis, even though an agent originally sold them

The previously given commission rates and contract terms minimize churn in several ways:

- Low up-front commissions (with good residual commissions) and a 180-day minimum reduce the incentives for subagents and independent dealers to churn customers.
- Residual commissions provide incentives to reduce churn, but, more importantly, make it impossible for the agent to leave the carrier.

To avoid channel conflict, the carrier (or reseller) and its agents should understand up front that the carrier will handle major accounts and direct marketing to consumers and small business. Its direct sales force will not cover any segment but major accounts. The carrier's direct sales force will not receive commissions for sales to accounts other than major accounts; if an agent sells a major account *occasionally*, the agent will receive an up-front but no residual commission and will turn the account over to the carrier.

No examples exist in which any market has been able to maintain any discipline on wireless telephone pricing. If there were some contract terms that could maintain price discipline between the carrier and its agents, it would be eroded by competitors or disallowed by law.

If agent contracts were nonexclusive, residual commissions would lose their importance in keeping talented agents and as an incentive to keep customers with one carrier. For maximum carrier performance under nonexclusivity, residual commissions would be dropped, up-front commissions would rise, and carriers would try to extend the minimum customer life to pay commission from 90 to 180 days to one year. Carriers who have tried to keep agents exclusive while offering no residual commissions and constantly reducing sales commissions have been left bewildered or suing their agents until they offered residual commissions.

Monthly performance information should be shared with exclusive agents freely as peers and partners of the carrier in the retail marketplace. The only information to be kept from agents is the performance of other agents and the retail performance of the carrier and resellers. The following typical information should be regularly shared with agents, and perhaps with direct dealers also:

1. Carrier activation and churn rates for the month (this allows the agent to tell if his or her performance parallels the overall carrier performance for the same period)
2. Churn
3. Customer satisfaction: sample taken once per quarter by the carrier
4. Breakdown of sales by market segments: by company size, management level of end user, industry, personal lifestyle of end user (income, location, automobile, etc.), and rate plan, which should be consistent with market segments
5. Usage per customer
6. Average bill size (ARPU)
7. Mix of subscriber pricing plans and average price per minute
8. Mix of brands of handsets and type (mobile, transportable, portable phones) once per quarter
9. Trends in slow payment and bad debt

Agent Financial Statements

Agents have no incentive to reveal their financial condition to their carrier or reseller. In many cases, agents fear that if they are financially healthy, then commissions may be reduced or support for the channel will be withheld. Relationships between agent and carrier may be strained because the carrier competes directly with its own agents, provides little marketing support, and/or attempts to lower costs through lower commissions. Giving the appearance of poverty, agents feel, may also discourage new agent entrants. At any rate, agents have few incentives to demonstrate financial health, especially where the relations with the carrier are not viewed as a partnership.

Contracts with agents should always require quarterly financial statements, although even these will always appear to demonstrate that the agent is not making any money. The carrier should review these to make sure that agents are successful, rather than to look for opportunities to lower commission costs. Only a long-term development of partnership and trust can permit the carrier and agent to function on terms of mutual support.

The illustrative agent income statement is developed on current performance assumptions in Table 6.3. Actual agent performance and characteristics can be used, like these, to develop a model that produces the income statement in Table 6.4, and these amounts can be easily estimated even without the agent's full cooperation. Such an analysis is useful for the prospective agent to understand the economics of his or her business, but also for the carrier to understand whether and how the agent or dealer will be sustainable through the programs the carrier maintains for them.

It is apparent that the fictional agent depicted in Table 6.3 utilizes equipment sales as a tool to sell service, rather than as a profit center, as the handsets are subsidized to an average price of $25. At the given sales rate, the agent would reach 7,000 subscribers in about $2\frac{1}{2}$ years. The residual commissions cover a large portion of fixed costs, while the current sales rate supports the variable portion of operations costs. Some sensitivity tests would show that the profitability of this agency is dependent on how well it can maintain a large customer base at low fixed cost. Thus, long-term profitability is not simply dependent on sales growth, but on keeping churn low so that sales are not absorbed by churn replacement. If churn is kept under control, residual commissions are pure profit.

If the sales rate and churn rate remain constant, this agent will reach a plateau at about 10,000 customers. Each year, the income statement will change drastically based on the relationship of the sales rate to the total customer base and the sales commission and residual commission rate that goes with each. Each agent, and all channels, must determine the optimum

TABLE 6.3 Agent Income Statement Assumptions

Item	Value
Sales rate per month	350
Churn rate per month	2.50%
General manager annual salary	$48,000
Sales staff number	10
Sales rate per salesperson per month	35
Carrier subscription commission rate	$300
Carrier residual commission rate	5.00%
Commissionable revenue per user per month	$60
Phone margin	−$75
Annual office salaries per person	$24,000
Rent and facilities per month	$3,500
Installations—percent of sales	5.00%
Installation price	$125
Phone wholesale price	$100
Installation labor	$40
Installation material	$50
Sales commission per sale to salesperson	$80
Marketing expense per sale	$20
Telecommunications expense per month	$1,000

TABLE 6.4   Agent Income Statement

Revenue	
Phone sales (350×12) at $25 each	$105,000
Installation	26,250
Commissions	
Activation	1,260,000
Residual at 7,000 subscribers	252,000
Total Revenue	$1,643,250
Cost of Goods Sold	
Equipment	$420,000
Installation	10,500
Total cost of goods sold	430,500
Gross margin	$1,212,750
Expenses	
Salaries and commissions	
Manager	48,000
Customer	72,000
Service (3)	
Technical, clerical, administrative	72,000
Commissions	336,000
Total salaries and commissions	$528,000
Rent and facilities	42,000
Marketing	84,000
Telecommunications	12,000
Installation labor	8,400
Total expenses	$674,400
Operating income	$538,350
Operating income as a percent of revenue	32.76%

long-term size of the operation that provides the best balance of the carrier's or reseller's quota demands and the overall objectives of the agent's business, for example:

1. Maximize gross profit
2. Maximise percentage return on sales
3. Maximize percentage return on investors' capital
4. Obtain the maximum manageable operation size
5. Maximize growth to position business for sale.

In summary, the agent channel originated with the first commercial cellular service and is still important to wireless telecommunications.

There are thousands of agents serving wireless service nationally. Since the need for technical and installation service is eroding, the channel is beginning to be displaced by retail and internal channels.

Trends

The current trend is for carriers to reduce commissions paid to agents and dealers and increasingly to convert their internal channels from a direct sales force for large accounts to carrier stores for consumer customers. Furthermore, carriers pay lower commissions for consumer customers versus business customers (based on which rate plan the customer chooses) and sidestep external channels to promote phone sales and service subscription direct to consumers.

Along with a decline in up-front commissions, the trend is for residual commissions to increase to as much as 10% of the revenue and to be paid as part of the compensation to internal channels as well as external channels. This recognizes the value of residual commissions as incentives for keeping customers and to ensure that the agent or salesperson can never leave the carrier without great income penalty as a loyalty incentive.

Carriers often pay up-front sales commissions conditionally; if a customer leaves the service before 90 days, the sales commission is taken back by crediting it against future commissions. These conditional periods are getting longer as revenue per customer is dropping and competitive churn is increasing, and carriers are extending them to 120 to 180 days.

EXAMPLES OF A WIRELESS CARRIER DISTRIBUTION CHANNEL MIX

BellSouth

1995: 11,000 retail outlets, 800 direct sales representatives, 175 major account representatives, 125 company stores

New programs—kiosks in supermarkets, MCI as a reseller, joint marketing with telecommunications service

1997: 10,000 total outlets, including 200 company stores, 4,400 direct sales representatives, 1,800 national retail outlets, telemarketing

Nearly 60% of new customers through captive channels in 1997. Commissions lower in internal channels

Selling wireline products in retail stores, "ready-made for bundled offers"

Size: 4.1 million customers (includes proportional share of customers in markets in which it has a minority interest), growing 14% in 1997 to 10.2% penetration

Omnitel—Italy

Customer base grew 1.6 million to 4 million in the first six months of 1998
While the base has grown 59%, business customers have grown 10% in the same period due to two new business rate plans. 90% of new subscriptions were prepaid rechargeable card customers
Acquisition costs under 60,000 lire per customer (1766L/US$)

AirTouch

2,000 distribution points including 100 company stores

AT&T

Has dealers, direct sales force, retail stores
Sales force has a national account services group for national sales
Operates a central clearinghouse for new national account orders and shipping cellular phones to national accounts
Recently added grocery stores to distribution channel.
In addition to company stores, added 20 "AT&T Wireless Express" phone centers in New York City
Implemented an automated kiosk, called Interactive Retail Information System (IRIS), which dispenses ready-to-use cellular phones to the customer and provides customer service features such as rate information, equipment prices, and local calling area maps (developed by Lightbridge, Inc., a cellular logistics company)

Bell Atlantic

Uses retail, sales agents, BAM Communications Stores, direct sales force, telemarketing, and resellers
Strategy is targeted distribution with different outlets for different customers
Direct sales force of 400 is being used for large accounts, BAM Communications Stores (120) (other reports show over 175), and authorized agents sell to the small business market

Retail served through Sears, Radio Shack, Circuit City, other retailers

In-house telemarketing handles promotional responses and is growing

Separate, specially trained sales force sells wireless data (CDPD)

Also sells through system integrators, value-added resellers (VARs), and business alliances

Volume discount for resale of CDPD

As of year end 1996 gets 60% of sales from direct channels (direct sales, stores, telemarketing) and 40% from retailers and agents

PCS PrimeCo

Has direct sales force for small, medium, and large businesses

Planning to end 1997 with 75 company stores

1,400 retailers

7

ADVERTISING AND PROMOTION

Advertising consists of any paid use of media to promote a product or company. It is usually included in the broader category of marketing communications, which also includes public relations and product and sales literature.

Promotion, as distinguished from advertising, is an incentive or offer specifically designed to bring customers to the final stages of the purchase decision. It may be presented through advertising of a specific offer, may be a public relations event, or may be offered through the salesperson only.

It is natural for the wireless carrier or reseller to introduce and promote the service through advertisements that develop the themes of the benefits of wireless service alone. But, like other intangibles, it is difficult to sell wireless as a service without the aid of tangible images of its product characteristics and benefits.

The provider of any product or service naturally believes that its products are the primary focus of a customer purchase decision rather than a component. Intel, for example, likes to believe that customers choose personal computers based on the fact that their chips are inside ("Intel Inside"[1] on the box). Likewise, wireless carriers believe that prospective wireless customers decide to have wireless service for all its benefits and need a wireless telephone in order to use it.

However, such is usually not the case. It has been demonstrated that the new customer's focus in wireless communications is often based on the

---

[1]Intel and "Intel Inside" are registered trademarks of Intel Corporation.

decisions regarding the features and price of the wireless telephone and that wireless service is a necessary but secondary consideration. While prospects need to go through the steps of exposure, awareness, etc., in their decision to become a wireless user, they almost always think of the telephone instrument as the primary component and "deliverer" of the service. They make their purchase decisions based on the choice of telephone, not the service provider. The customer will most often subscribe to the wireless service provider associated with the seller of the phone (see Chapter 2, "Product Attributes").

In recent years, there has been an increasing awareness of the differences between carriers, especially with the addition of PCS and E-SMR in most cities. Today there is a mix of advertising for the service and the phone, but the choice of phones and the subsidized low price of the phone are still the most frequent basis of promotions.

Thus any sales channel will often get the business for the carrier based on selling the customer the phone. Resellers and independent dealers who are not aligned with one carrier will be able to influence their phone customers to subscribe to whichever carrier they recommend. This has important implications for channel strategy as well as promotional strategy. The carrier cannot control sales channels by limiting who will be allowed to promote the service; instead, it must attract the channels that are most successful at selling telephones, the focus of customer attention.

In order to understand this phenomenon properly, I like to use the analogy of the paintbrush manufacturer. When it comes to painting, there is nothing more important, in terms of productivity or the quality of the final product, than the paintbrush. Certainly, paintbrush manufacturers are proud of the quality of the paintbrushes they produce and would like potential customers to know it. However, no matter what they might do to convince people that their paintbrushes are superior, that people should seek out stores that sell them, and that they have the best variety and price available, people never make the paintbrush the focus of their painting job decisions. People rarely even consider paintbrushes when painting except as an afterthought, unless they are professional painters. Despite the importance and rationale for using the right brush, people will always begin a painting project by selecting a color. They may remember the paintbrush just before they walk out the door. Consumer behavior may be irrational, but it is extremely difficult to change.

Similarly, whereas in some cases customers may actually make their purchase decision based on wireless service, the service provider must remember that the customer most often will come to purchase wireless service first being brought to the decision through the vehicle of the tangible, visual wireless phone. It may not be rational nor practical, but it is virtually impossible to change. It is easy to work with these assumptions as long as we do not let our ego get in the way and try to force the customer to be attracted to wireless communications by the service.

There are three lessons about the promotion of wireless service through the attractiveness of the telephone:

1. Customers believe they have obtained "wireless" service wherever and whenever they buy a wireless phone, not just where the carrier's branded service is sold or its authorized sales outlets are located.
2. The customer associates wireless service with visual images of the wireless phone.
3. The consideration and thus the offer of wireless service are subordinate, secondary, and after the sale of the phone.

Advertising serves to attract people to consider wireless service, as well as to position them against competitors. However, while it is tempting to position the customer using the benefits of wireless *service*, wireless service *advertisements are not successful unless they portray some pictorial aspect of the wireless telephone instrument*. Without the telephone, the customer cannot visualize wireless communications effectively in an advertisement.

The need to portray the telephone in advertising applies both to initial educational advertising at the introduction of wireless service and to competitive and promotional advertising as the market begins to grow and mature. Wireless markets appeared to reach a point early when the base of customers was large enough to cause word-of-mouth referrals to become a more powerful stimulator of interest in wireless service than any combination of advertising and public relations work. All wireless markets have long since passed this point, when penetration reached $\frac{1}{2}$% to 1% of the market.

At this point, wireless telecommunications has credibility as a valuable service in the market. Educational aspects of advertising cease to be an incentive; early adopters have mostly accepted the new service and are opinion leaders for their peers. As we alluded to in the chapter on market segments, potential customers learn of the benefits of wireless service and are drawn to the first stages of adoption—exposure and awareness—through the following.

1. Interaction with opinion leaders among their peer group or business associates who are already using wireless
2. Increasing frequency of receiving calls from people who, based on the background noise associated with the outdoors or with a moving vehicle, are obviously using wireless
3. Increasing frequency of receiving calls from people who compulsively *tell* them they are calling from a wireless phone (but sometimes this is

an explanation for the quality of the reception or the need to keep the call short; often it is merely a power trip)

4. Being a passenger in a car equipped with a wireless phone
5. Increasing frequency of spotting wireless users on the street or a wireless's characteristic "pigtail" antenna on a significant number of automobiles.

Now that all wireless markets have reached this stage, customers use advertising to make a trial purchase decision rather than merely to understand wireless and become interested in it. As far as awareness is concerned, the medium becomes the message. The customer who has not thus far become aware of wireless communications as a benefit may not always notice the content of the advertising but notices how many advertisements there are and then realizes that the concept is for everyone (perhaps including him or her), rather than a fad for a special few. Another benefit of encouraging multiple sales channels is the appearance of multiple advertisements from different providers in multiple media, each getting to different kinds of customers with different promotional messages.

Since the population is generally aware of wireless telecommunications, the content of successful advertising concentrates on the trial purchase decision, focuses on the telephone or promotional offer rather than the service characteristics, and provides "ease of entry" through low, time-limited price offers, differentiating the features of various telephones, or through free premiums offered with the purchase of a phone (CD player, extended warranty, speaker phone, or other accessories). Certain types of media, such as television, may lend themselves to communicating the competitive superiority of the service rather than promotional offers.

The customer is brought to the decision to purchase through the avenue of a choice among attractive offers, whether conscious or not. The sales channels need such offers as one method of producing qualified sales leads and attracting customers to retail points of sale, although leads generated solely through price offers usually attract customers who are price-sensitive and already predisposed to buy. The direct sales force should be directed at those business customers who comprise a target market with a high need for wireless service who need to be convinced and are not price sensitive.

Another feature of advertising that recommends a time-limited, specific offer of price, free minutes of use, or a free premium is its measurability. The success or failure of the advertising can be easily measured through the sales success with prospects who were led to the sales presentation through the advertisement.

The directness of the promotional offer is an important aspect of its immediacy and effectiveness. By directness we mean the number of steps involved between response and purchase. Media advertising or a direct mailing that invites a customer to return a card or call to obtain a brochure introduces a first step. The customer then receives a mailer or brochure containing service information or a specific offer. This may invite the customer to call or return a card as a second step. Often, a follow-up call from sales removes the requirement for the prospect's initiative for the second step.

The laws of diminishing returns apply severely to such programs. If the response rate for a mailing averages 2% (high), the response from a two-step process is at best 0.02×0.02, or 0.4%.

Even if a customer responds to an advertisement by requesting to subscribe or to have a salesperson call on him or her, a requirement to have someone call them back constitutes a two-step process and diminishing returns. Anyone handling calls in response to an advertisement should be able to initiate service or set an appointment for a sales call rather than have someone call them back. A "live" handoff to a salesperson or customer service representative rather than a "warm" lead through a return call is mandatory.

The advertising content must not only provide a direct response mechanism, but a unique code identifying the offer and the medium for tracking purposes, as well as a time-limited offer (e.g., "before June 30"; "for 30 days only") to urge a call to action. Only a single channel should be offered for response, although it may have several dealers or convenient locations within a single channel or a single call to locate the nearest one. Advertisements that invite customers to call the carrier, see the nearest agent, and call for the nearest dealer are not only unfocused but difficult to measure. Printed advertisements often include a mail-in reply option. However, over 90% of replies are by telephone, and these usually reflect higher immediate interest in immediate purchase. Mail-in replies reflect "softer" interest and are usually superfluous.

In the same manner, advertising copy must be limited to one offer and to one customer segment, as well as through one channel. It is tempting to extol the virtues of wireless service, illustrate several applications and the attractiveness of several offers, and emphasize the quality of service offered at multiple locations in one advertisement. However, the desire to overwhelm the customer with the superiority of the carrier's overall offering through multiple messages merely diffuses the importance of each message. The stimulation of interest is much more effective through the exposure of the potential customer to multiple different offers, in different media, through different channels over a short time period, each promoting one benefit, one offer, at one place. This occurs partly through good management and partly through serendipity when multiple channels are used. On any day the potential customer might see an advertisement promoting the price of several wireless telephones, another promoting the features of one manufacturer's newest unit, another stressing quality, another emphasizing the importance of doing business with a local independent small dealer, and another with a free premium.

In the mature market, more and more wireless users become repeat buyers, and their knowledge and experience permits them to discern the differences among carriers more carefully. These customers are often enticed by offers of better service quality and free air time rather than phone prices. However, they are often attracted by the features of a new phone or an upgrade from analog to digital service through the phone, and they still require the visual image of the phone. Promotions of free air time merely cause customers to switch from one carrier or provider to another and do not attract good, new customers. Repeat buyers who need an additional phone or upgrade will be attracted to the same promotions that attract first-time buyers. They will go back to where they were treated well and shun carriers and dealers who have treated them poorly.

MEDIA

As mentioned, the highest-potential target market for wireless service is still the mobile small-business executive. However, as individual markets and the wireless concept mature, much broader consumer mass-market segments are attracted, and most of the business users already are subscribers. To reach the desired market, media must be targeted as well as the advertising message. This not only efficiently directs the advertising to the target audience but also controls marketing costs. Television, for example, as a mass-market medium, is not usually appropriate until the market is mature. Television is not only expensive in media costs but also in production. This is evidenced by the low quality of local television commercials, usually "starring" the proprietor. But television can be effective in promoting the brand and the competitive advantages of the service, while other media—radio and newspapers— promote time-limited special offers.

The mobile small-business executive and the upscale consumer are appropriately addressed by radio in the automobile during commute hours ("drive time"), especially in conjunction with business news and traffic reports. Advertising in the newspaper used to be most effective in the business and sports sections, concentrating on male business executives. Today, advertising is usually found in the first news section.

A mix of alternating newspaper and radio advertising directed at the same offer and/or benefit reinforce each other for a particular campaign and is usually most effective. Experience will determine the appropriate days of the week for newspaper advertising (usually the beginning of the work week for wireless services) and the best stations and times for radio. As markets have matured, the advertising and promotional messages have expanded to include more general sections of the newspaper and more radio stations with a variety of formats preferred by the broader consumer segments.

Television has been adopted in some cases by carriers in mature, major markets in which consumer segments are being significantly penetrated and carrier competition is keen. The effectiveness of television must be measured even more carefully than other media. With television advertising directed at consumers, we encounter a phenomenon in which the most expensive advertising medium is being used to attract the least valuable customer in terms of monthly revenue per user, customer longevity, and customer loyalty.

PUBLIC RELATIONS

Wireless marketing has always been able to capture media attention through its novel applications, high-technology interest, and public interest angles. Wireless telecommunications has increased highway safety, traffic reporting accuracy, and crime prevention, to name a few public benefits. A good public relations person can supplement advertising and word-of-mouth public awareness with frequent news releases and quotes mentioning the company name concerning new technology, coverage additions, wireless use by public agencies, etc.

The carrier must take the responsibility to recognize the large role of wireless communication to improve public services and the public good and to contribute to interests such as emergency communications plans and other public needs that wireless communication can serve well. While the carrier's responsibility to promote the community interest should be at the forefront of the carrier's participation, the resulting good publicity can be an excellent supplement to the marketing program.

Specific community interest programs can provide important services, highlight the wide applications of wireless service, and help to make the company a good corporate citizen. Many carriers provide wireless and other services to local charities. Some may have special wireless applications, such as telephone services for victims of spousal abuse.

ADVERTISING PITFALLS

There are several pitfalls to avoid when advertising, some of which apply uniquely to wireless services.

1. *The "burning eyes" syndrome.* It is easy to fall into the trap of advertising heavily in response to highly visible advertising of competitors. The name comes from the way you react when you or your boss sees such advertising, which may demean your own offering, make inflated claims about your competition, or copy your strategy. Resources may be thrown at creating a countercampaign at least as visible and expensive as the one that stirs your angry emotional response. This can cause a wireless provider to disrupt its entire annual advertising and promotion strategy and budget for no reason. This type of advertising does not communicate to potential customers and merely consists of carriers arguing in public. Advertising must be strictly guided by objectives, schedules, and strategies in the annual marketing plan.

2. *Advertising Measurements.* The measurement of advertising recall, brand awareness, and other aspects of the advertising program is normally effective in the market for retail consumer goods in which there is no direct measurement of the effectiveness of an advertisement in producing a sale. If most wireless advertising is correctly aimed at sales generation, such measurements are subordinate and unnecessary compared to direct measurements of inquiries in response to specific offers.

3. *The desire for visibility.* All advertising attempts to "move out of the clutter" with larger, more visible newspaper advertising; more frequent, creative, and varied radio advertising; and the visual and leadership appeal of television. By measuring effectiveness rather than serving these desires, the advertising program will better achieve its goals.

4. *Overmanagement.* Everyone in the company will want to approve advertising and recommend advertising strategies. In addition, everyone in the company will criticize advertising based on its appeal to his or her personal tastes. It is important to remember to be objective and concentrate on the tastes and response of the target market. It is very important for senior executives to stay away from the temptation to provide personal input on advertising and to insulate marketing from criticism from other departments, yet be merciless in the demand for improvements in objective measurements of results.

Finally, all marketing managers and corporate executives are cautioned to discipline themselves carefully to refrain from what I call "auto-market research," the projection of personal preferences on the market as a whole. This is a condition in which managers assume that their own tastes are congruent with those of the target market and substitute such opinions for objective research. This results in acceptance of new advertising programs and new telephones from manufacturers, for example, without a proper evaluation by potential customers.

Always rely on market research and related focus group reaction to advertising and promotions, as well as service features, impressions of customer service, sales presentation quality, and all parts of the marketing process. Do not rely on your own personal judgment to determine whether advertising is "good" or not. Never ask your boss for an evaluation of a program for which the targeted audience is the better judge.

8

MARKETING MEASUREMENT

As with any good marketing program, one of the most important aspects of wireless marketing is the measurement of marketing efficiency and effectiveness. Whereas most will admit its importance, the implementation of a disciplined program to (1) measure marketing effectiveness and (2) act on it is rare.

There are many components of marketing effectiveness to measure:

1. Revenue versus sales expense
2. Customer gain versus sales expense
3. Revenue versus marketing expense
4. Customer gain versus marketing expense
5. Store traffic or lead generation versus advertising expense
6. Customer satisfaction
7. Customer churn

This list is not exhaustive, and many more elements and combinations might be included. However, this list will serve to illustrate the meaning of various measurements and how they are applied to decisions and marketing actions. Let us look at each one more closely.

1. Revenue versus sales expense is an important measure but has little to do with marketing effectiveness in a service environment. In a manufacturing

environment, revenue is achieved through sales booked during the current period. Thus there is a direct relationship between current sales expense and current revenue. In a subscriber service, however, the sales effort results in future revenue over the life of the customer (the length of time the customer subscribes). Therefore the measurement of sales expense versus revenue is an important diagnostic of current financial performance but does not measure sales performance.

2. Customer gain versus sales expense is an important measure but often is given as *net gain* (new additions minus churn) instead of *gross gain*, the actual number of new customers sold during the period. It is measured as net gain because, from the company perspective, net growth is the bottom line for company growth and financial gain. Also, it prevents sales channels from inflating results by activating "imaginary" customers who are dropped after a short time. Actually, the sales channels have virtually no control over churn unless they are directed at the wrong customers and should be judged only on sales. Churn is a problem to be solved through customer service and loyalty programs.

3. and 4. These are analogous to components 1 and 2 for marketing instead of sales. While it is good to measure marketing and sales separately, there should also be a measure of them combined. The two areas may be organizationally separated but are really a single function: sales is one portion of the marketing mix.

5. The store traffic or lead generation versus advertising expense measure recognizes that advertising must be directly evaluated for results. Optimally, the actual sales that result should also be measured. While much of the sales rate may be outside the control of advertising pull, the advertising may be pulling customers that are in the wrong segment, unusually price-sensitive, for example. Thus even a promotion that provides excellent store traffic and lead generation for direct sales may not provide sales results comparable to other promotions that provide fewer or lower-quality traffic and leads, but better sales results.

6. See the detailed section on customer satisfaction later.

7. Churn is an excellent example of a measure that everyone watches and worries about, but about which little is done. Most of the time, it is thrown on the shoulders of sales, as mentioned previously, to be responsible for net gain instead of just gross sales, even though sales has no influence on churn. Sometimes it is used as a customer quality measure, although no one has been able to show that churn is substantially reduced with better customer care. Unless churn is over 2.0% per month, probably little can be done to improve it as it would not show up in customer satisfaction measurements. If it is under 1.5%, chances are the carrier is not actively pursuing new market

segments aggressively, and may be a sign of marketing and sales complacency rather than a positive measure of control.

Providers with low churn will say that their customer care program is at the root of it, but this response is more public relations than cause and effect. While various measures should be taken to control churn when it is caused by competitive pressure or customer dissatisfaction, providers should understand that, in general, only the churn above 1.5% to 2.0% per month can be controlled. As providers continue to push subscription by recruiting market segments with marginal need for wireless service, churn will rise. Because of the increased emphasis on churn, especially in mature markets, there is a separate chapter on churn and customer loyalty (Chap. 14).

There are three basic rules for measuring results in wireless marketing; they apply to market research as well as marketing and sales performance.

1. Follow Steuernagel's rule: "What would you do if you knew?" Before you measure any parameter, think of the wildest range of values it could possibly assume and the action the company would take in each case. If you cannot imagine the company taking action or if management arbitrarily does what it pleases, do not spend the time and money to measure it. As an example, if overall customer satisfaction dropped from 90% to 80% (the absolute bottom), would you do anything to correct it? Would your research tell you what area you need to correct?

2. Measure only the parameters that are controllable; measure organizations based on the things they can control. Gross customer gain is controllable by sales and marketing; net gain is not. Sales leads generated by advertising are controllable by marketing, but closed sales resulting from advertising are controllable only by the combined marketing and sales team.

3. Avoid the temptation to measure things because they are measurable. Revenue per customer is measurable but not really controllable (unless you restrict marketing to high-usage segments, in which case total revenue is reduced). On the other hand, total revenue is measurable and controllable. Sales cost as a percent of current revenue is measurable and useful for finance to measure this month's cost performance, but it is not useful to measure sales performance. No amount of sales effort will significantly raise this month's revenue in the wireless environment. Cutting sales cost will lower it as a percent of revenue for the month, but this measure will rise severely due to reduced revenue in the future if sales effort is lowered today.

Carriers often use measures of the success of sales force management instead of whether the right channel is going after the right target. The carrier measures sales per salesperson, cost per sale, and sales per store. Rather than compare one channel to another, the carrier tries to improve each number from one month to the next by managing the measure and the people. Retailers and dealers employ similar measures, using the statistics of individual salespeople and subagents to chastise them into doing better, rather than aggregate measures.

Carriers often do not *want* to know the cost of their direct sales and store channel compared to other channels because they want to control sales and do not want any reasons not to.

USEFUL MEASURES

There are two basic areas of understanding the measurement of marketing success. *Effectiveness* is the degree to which the marketing process—advertising, sales channels, etc.—is successful in reaching all of the targeted potential market and converting it to customers at a rate consistent with projections. *Efficiency* is the measure of long-term revenue versus the total cost to get and keep the customer—the ratio of unit output to input. Often, marketing is responsible for effectiveness while staying reasonably within budget. Without diminishing the importance of effectiveness, efficiency is the source of improvements in effectiveness and profitability, because it frees dollars for the resources to be more effective.

An organization that gets only one new wireless customer but at a cost of under $100 is remarkably efficient, but not very effective. An organization that exceeds customer gain projections by 20% is very effective, but not very efficient if each new customer carries a marketing or sales cost of $500+ and a low monthly usage rate.

There really is no substitute for sales and sales growth in relation to cost as an overall marketing measure for wireless carriers. All other measures are subordinate indicators of problems.

Some additional measures that are useful, controllable, and "actionable" include:

Gross sales per salesperson (versus peers in the same channel)

Gross customer gain

Increase in customer gain

Increase in revenue

Leads and sales generated by promotional campaigns (versus previous similar campaigns and versus expectations)

Revenue per customer over the average life of a customer in each market segment and channel (expected revenue per customer)

Expected revenue per customer divided by the sales and marketing cost per gross customer gain (by channel).

KEY PERFORMANCE INDICATORS AND BENCHMARKS

Performance measurement should cover key areas to ensure that operations oversight maintains levels of service at or above the best practices of the industry. Some of these key performance indicators (KPI's) and levels maintained by industry leaders are given in Table 8.1.

In addition to the benchmarks, next to each the operator should list the current level and the goal, as indicated. Depending on company goals and satisfaction studies showing customer priorities, various KPI's may have very challenging goals or else are set to approach the benchmark in stages if the current level is far below the benchmark.

Note that the only network performance indicators are dropped calls and blocked calls. The many complicated and necessary measurements for

TABLE 8.1 Key Performance Indicators and Levels for Industry Leaders

Key Performance Indicator	Current Level	Benchmark	Goal
Marketing and Sales			
Acquisition cost per customer (except resale)		< $300	
Sales productivity			
Direct salesperson		> 35 per month	
Company store		> 300 per month	
Customer Service			
Speed of answer		< 30 seconds	
Abandon rate		< 5%	
One-call resolution		> 90%	
Retention unit save rate		> 40%	
Customer Satisfaction			
Customer satisfaction index		> 90%	
Customer life revenue per acquisition cost		> 6.0	
Churn per month		< 2%	
Dropped calls		< 4%	
Blocked calls		< 2%	

network operations and other departments are necessary to manage their function. But most can be summarized for the purposes of customer satisfaction in very few areas. Billing accuracy, for example, is included in customer satisfaction and can be further investigated in more detailed reports. Since network service quality is a high-priority specific concern for customers, it is included in this list as the two symptoms customers actually experience and report. Coverage is also a network parameter of specific interest to customers but is not a performance issue.

GENERAL MEASURES OF MARKETING SUCCESS

The wireless provider must have a few measurements of immediate performance that are available daily or weekly. There are really only two categories: customers and usage. Usage is more important than customers; however, not much can be done about it. (I have always tried to correlate periods of high wireless usage with news items and other events to try to understand what causes high usage and what can be done to keep usage high. I have found that there are only two items I need to create to ensure high wireless usage: traffic accidents and bad weather.)

It is possible to increase usage with new applications such as voice messaging and information services. This would be unnecessary if only high-need, high-usage customers were attracted to wireless communications. Concentrating on high-usage customers gives more immediate results and leads faster to higher revenue. Unfortunately, too often we are given incentives to increase customer counts indiscriminately for the benefit of investment analysts and our superiors. We are then asked to apply ourselves to the challenging task of making users out of them.

A report of usage levels may be necessary, but as mentioned usage levels are not easily controllable. The most basic measure of marketing and sales for diagnostic purposes is gross customer gain. In the morning results should be available from the previous day so that appropriate adjustments can be made to advertising, lead-generation efforts, sales force levels, commissions, etc. These tactical adjustments are most important but cannot take us away from the longer-term tasks of intelligent planning, measurement, and strategic decisions.

Remember that having instant information is only important if you have the capability to make instant decisions and take quick action. Gather only the information that enables tactical decisions that the recipient is allowed to make and in the time frame permitted by the organization.

Carriers generally pay far too much attention to customer gain alone, especially as revenue per customer decreases. Perhaps the only measure more useless as a gross measure is the potential market as expressed by market population ("pops") used for investments in wireless markets. Direct attention to revenue gain through additional services, usage, and price increases, as well as customer gain, are all important. Too much emphasis on customer gain will force marketing to attract more marginal customers with low usage at higher sales costs per customer.

MEASURING PROMOTION SUCCESS

The primary, short-term need during a promotional campaign is to measure the effectiveness of a particular advertising or promotional campaign by monitoring immediate response on an hour-by-hour or day-by-day basis. The promotion must be designed to have a specific order form, telephone response number, customer service team, mail-order box number, or whatever means it takes to identify responses to particular programs. Additional information, such as time of day and person responding, may be appropriate depending on the ability of the organization to take tactical advantage of more detailed information or to move more quickly. If multiple media are used, the media must be identified if other information is the same.

The identification of the demographics of the customer is just as important as the information collected in order to complete the sale. However, the collection of information from the customer must never interfere with the sale. You can probably only find out three or four characteristics of the customer before the questioning becomes intrusive; make sure you do not have to ask them where they learned about the offer: the means of reply should tell you that.

The importance of tracking a promotion is not just to measure its success; it collects prospects for future promotions if customers do not buy, for example. But direct marketing is almost synonymous with database marketing. This is the use of a database not only to collect sales information but to collect information on the success of the promotion as feedback into an ever-improving cycle, an "iterative enhancement" of the direct marketing process. Even if the promotion is successful, careful measurement of the results will tell you whether or not the promotion hit the target market. If it missed but was still successful, then you have hit another market and the original target is still unpenetrated.

If the promotion was not successful, the data collected should tell you whether the target was missed or the offer was wrong. Either way, each successive promotion should be much more productive based on careful tracking of the responses of different target markets to different types of promotions.

While the direct marketing (telemarketing, direct mail, catalog, database) aspects of promotion are important for both lead generation and closing, instruction about all aspects of these topics is beyond our scope. Good books on direct marketing will help with technique; our purpose here is merely to recommend that some techniques of direct marketing and database marketing are appropriate for measuring and improving promotions.

CUSTOMER SATISFACTION

Customer satisfaction studies are important as a measure of quality and the various components of customer care that, if neglected, can lead to increased churn, low usage, and lack of adoption of the service into the person's lifestyle over the first few months as a wireless subscriber. Such question-naires can also include questions about demographics, which should verify data obtained by other means about the preferences and proportions of customers in each market segment.

Customer satisfaction studies are also very easily manipulated and should always be conducted by an outside market research firm, as with all market research. When properly conducted, customer satisfaction can be used as a performance measure if it can be correlated with long-term improvements in revenue per customer and customer loyalty. It is also very easy to produce a customer satisfaction study in which >90% of customers rate overall service as satisfactory or better. What is really important are the sources of dissatisfaction for the remaining <10% and statistical confidence that satisfaction is improving.

In order to provoke action, customer satisfaction must be broken down into the sources of satisfaction and dissatisfaction in customer questionnaires, so that specific sources of dissatisfaction can be corrected.

Typical questions for the customer satisfaction study would include (rated on a 4 to 5 point scale of excellent, good, average, below average, poor or similar):

Overall satisfaction
Ability to access system
Billing detail and clarity
Size of service area (and areas of poor service)
Dropped or unsuccessful calls over the past 30 days
Interactions with customer service staff, if any
Wireless service as a good value for the money
Open-ended comments

Other questions would include experience with salespeople and technical people. Too often, the content may get bogged down in questions about whether the salesperson gave the customer instruction or information in particular areas; this often occurs as an overreaction to learning of some customers being misinformed about prices, operation, etc. This results in the company requiring salespeople to inform customers to the point of distraction about things customers do not want to know. Asking a customer if he or she was informed about roaming procedures does not tell you whether they wanted to hear about it. It is bad enough to make a customer a victim of your attempts to discipline salespeople; do not make him or her the scorekeeper also.

A customer satisfaction study, as shown by the previous sample questions, applies to other areas besides sales: coverage, service quality, etc. The study should be conducted on a regular basis, using the same questions and same scoring, so that changes can be fairly represented. The market research organization can show you how to conduct a *longitudinal* study, in which the same customers are checked for changes in attitude over time. This should be considered in addition to ongoing studies of different, random samples and special studies of important subgroups of customers.

Another pitfall of customer satisfaction studies is redesign. Results for a longitudinal study may be manipulated to show slow but steady improvement between 80% and 95% satisfaction. When improvement stagnates, the study is redesigned and the results designed to fall back to 80% with the excuse that the redesign caused the drop. Then results creep slowly back up to 95% before another redesign.

As with other areas of measurement, two reminders are needed:

1. Do not study anything you are not willing to change; examine the range of "What would you do if you knew?" before embarking on a study.
2. All marketing, sales and customer satisfaction measurements are subordinate to financial results. It must be shown that any improvements in marketing measurements will lead to improved financials.

Finally, remember the financial responsibility of marketing is to keep costs down as well as revenue up. Ratios of life cycle revenue to sales cost are

TABLE 8.3   Cost per Activation: Retail Channel

	Agent	Carrier
Sales cost	$250	$350
Including marketing	$400	$550

important to determine relative efficiency, but raw costs per activation like the typical ones in Table 8.3 are much more urgent in their message that costs are too high in a business in which revenue per customer is eroding.

It can be tempting to substitute an inexpensive channel for an expensive one. However, this is not practical. You cannot find more successful consumer electronics retailers in a market if they appear to be the most efficient. And you cannot get customers from one segment to buy through a channel with which they are uncomfortable.

9

COMPETITIVE BEHAVIOR AND ANALYSIS

Many easy analogies of marketing competition are made that liken competition to war, to a battleground with fight-to-the-death combat, to winning and losing. These are important aspects of competition. The portion of competitive thinking that assumes that every sale the competitor makes is one that you could have had is an important part of the aggressiveness and attention that competitive vigilance requires.

At the sales level, competition is most often a head-to-head bid for the same business from the same customer. There are several aspects of competitive behavior, however, to which good marketing techniques can contribute to make competitive superiority easier and less risky.

COMPETITIVE DIFFERENTIATION

The competitive threat is worse when all competitors offer the same product at the same price to the same customers. It is important to differentiate wireless service from the offerings of all others in the marketplace. But even differentiation means little for marketing purposes unless the customer can see it. This is done in the following ways, for example.

1. The product can be distinguished and differentiated by the features it includes, even though it is intangible—the size of the service area, inclusion

113

of vertical features and/or voice messaging, the refund policy on dropped calls, the information services such as traffic and financial reports, etc.

2. The product feature does not exist as a distinguishing factor unless it is promoted heavily. Make sure you are willing to concentrate on these factors in advertising (other than handset and air-time promotions), literature, public relations displays, and sales presentations. The value of a distinguishing factor as a promotional item can far outweigh its real importance to users. Fiber optic trunks, for example, do not by themselves have significance to user benefits but are an excellent distinguishing product feature to publicize if the other carrier does not employ them, as a perceptual indicator of the carrier' dedication to quality and the latest technology. Sprint, of course, has done this with its long-distance network.

3. Pricing is the most visible distinctive feature of a provider. By having a rational set of rate plans with meaningful names that relate to customer groups and benefits that can be easily distinguished, much competitive advantage can be gained for cases in which the competitor is not as astute in pricing.

4. In addition to distinctive messages, promotions must have a distinctive look that is the trademark of the provider as much as the logo. The announcer's voice, a familiar spokesperson, the size and placement of advertisements in newspapers, colors, types of pictures and fonts, etc., all contribute to a certain "look". This does not mean that whatever you do first must stay. When you find a good look, you will know by the compliments. Then stay with it for as long as it works.

5. The quality of your salespeople, internal and external, is important, but the difference should also be identified with unique symbols. Colored blazers, large logos, a distinct color scheme for all store and agent locations, etc., tie together the quality with common visible reminders and show how ubiquitous and uniform the team is.

6. Most carriers have missed the point of branding. Wireline services use their company name because they believe it is important, and it is. But carriers, wireline and nonwireline alike, have missed the opportunity of branding their retail service separately from their carrier service. Just as Maytag Corporation sells washing machines at wholesale generically to Sears as an Original Equipment Manufacturer (OEM) supplier and under their own brand name through different outlets, carriers can use the corporate name as the generic name of the service to make the most of the corporate reputation and a separate name for premium retail services.

While many carriers have given brand names to their information services and voice messaging services, they have failed to develop a brand strategy that distinguishes all the benefits of their retail offering from the carrier offering. The retail service differs in the quality of its sales representatives, customer service, billing, and pricing packages at the least. The reseller often does not want to refer to the carrier's name, despite the amount of equity it has in the market. AirTouch Cellular, as one example of good brand differentiation, has successfully branded their digital service as Powerband^{sm}; more carriers are starting to show some ability to distinguish parts of their retail offering through branding, especially with pricing.

Even a nationally franchised trademarked name, such as Cellular One$^{®}$, can be used to distinguish all these other aspects of premium retail service separately from the carrier name. In some cases, the name of a reseller such as Motorola can be stronger than the carrier's own company name, and the Cellular One name might be better used to brand retail aspects of the service. Creation of a separate name increases the apparent number of providers in the market. However, many carriers do not even recognize that there is a retail offering to distinguish from the wholesale carrier offering shared with resellers. When wireline carriers come up with a new nationally franchised name, perhaps they will know what to do with it.

BOUTIQUE VERSUS INDUSTRY

Competition is not necessarily a war on all fronts. The existence of competitors in a marketplace shows that the product is of interest to the mass market rather than a speciality item for a narrow set of users. Competition provides a broader distribution, multiple offering choices, local service, and competitive prices for the product and its associated accessories to more market segments. A wireless market in which several carriers "own" wireless services by maintaining their own internal sales as the primary channels may be too careful at keeping competition out. In such a market, wireless appears to be a "niche" or specialty item intended for a very narrow market, available in only several flavors and only in a few locations that only those who seek them out can find.

By encouraging multiple channels to participate in the market, wireless communications becomes a ubiquitous presence: in multiple newspaper advertisements every day, in window stickers at appliance stores, in a public relations event held by a reseller, in an open house held by a broker-agent for dealers. Instead of several carriers battling for first place using their own exclusive channels, there may be dozens of dealers, resellers, and agents competing every day—and loving it—thus keeping sales costs and handset prices low without the carriers receiving the scars.

To the customer, the existence of multiple end-user providers, some offering poor quality and low prices and others a different mix, gives the appearance of a healthy industry and a product with mass-market appeal rather than a niche product for someone else.

Thus, while wireless carriers must compete aggressively at the retail level using their own sales forces and stores, they must support a competitive environment that enables dealers, resellers, and other external sales channels to be successful, in order to allow the fastest and largest long-term penetration of wireless communications into the marketplace; in short, carriers must work to make wireless an industry rather than a specialty within each local market.

COMPETITIVE ANALYSIS

The first step in being competitive is to understand your competitor. Competitive analysis is an ongoing function, especially in tactical adjustments to the daily sales and promotion activities of the competition, but is also an important component of the annual and strategic plans and must address strategic differences and the current status of the company versus the competition in all areas. Competitive analysis is not an event for managers to show their supervisors how hard they are trying to fight competition but an honest evaluation of the facts. What are the feature advantages of their systems? How well are they controlling call quality and system capacity? How is their distribution organized in size, quality, and location? How good are their advertising agencies? What are their marketing and advertising budgets? What happens when you call customer service? What are the revenues and market shares of the other carriers and all players in the market? What segments are they targeting?

The answers to questions such as these must be laid out side by side with your own operations to determine the area in which each competitor is the strongest. This analysis is the prime determinant of your positioning strategy based on strengths that are difficult for competitors to match, as well as competitive strategies to meet competitors' strengths and strategies. These strategies need not directly attack competitors. You can attack the competition head-on, drown them with your own unique strategy, and move the competitive battle to a different focus, or sidestep them by concentrating on a different segment, channel, or marketing element.

UNIQUE CUSTOMER PROPOSITION

While we have outlined the advantages of time-limited offerings to new customers based on handset-related promotions, the overall advantages of the company and the carrier over the competition must be portrayed to all local industry players as well as to prospective and existing customers. Rather than use periodic advertisements featuring various aspects of the service, the carrier should adopt a unique customer proposition that differentiates the service continually in all corporate communications and with which customers can easily identify as a valuable, real, and distinguishing benefit of one carrier over another.

Tandem[1] for example, in the computer industry, has positioned itself as the leader in applications requiring low downtime, adopting its name from providing computer power in tandem for reliability and adopting its product line name and overall theme as "NonStop" computing, which has survived years of competition as a virtually unassailable position in the market.

This unique proposition should be tagged to all announcements, advertisements, billing, and literature, in the form of a short motto or statement attached to the company or service name, appearing in all communications. It refers to a feature that relies on the competitive strength of the company, which cannot easily be duplicated or compromised. Statements as to leadership, size, and length of service do not connote a benefit, for example, "Alpha Cellular ... the leader in wireless service in the Metro area." Rather, statements should be positioned based on long-term advantages and benefits, as the examples given here demonstrate.

Advantage: For superior coverage:

Statement: "Alpha Cellular ... Wide-area wireless service"

Advantage: For a carrier that is leapfrogging the competition in new system development:

Statement: "Alpha Cellular ... The new generation of wireless technology"

Advantage: For the carrier with fewer service interruptions or dropped calls:

Statement: "Alpha Cellular ... Guaranteed nonstop wireless service"

Advantage: For the carrier that is dedicated to staying ahead of demand in system capacity:

Statement: "Alpha Cellular ... The Instant Access Company"

---

[1]Tandem and NonStop are trademarks of Tandem Computers Incorporated.

Cellular One, for example, uses the service mark "Clear Across America[sm]." Note that these examples are benefits that are not specific to the internal sales capability or the retail customer service operation but apply to broader benefits of the carrier operation.

COMPETING ON YOUR OWN TURF

The carrier has to be careful to avoid competing on the competitor's terms. This can occur very easily as a result of a successful promotion by a competitor or one that is very visible in the marketplace (see the "burning eyes" syndrome under in the section titled "Advertising Pitfalls" in Chap. 7). A program or communications strategy that is similar to the competition, defensive of the company's own attributes in the same area ("We have wide coverage too!"), or otherwise related to the competition's message automatically positions you as a follower or secondary player in the marketplace. Instead, you need to find a successful way to communicate a unique program that emphasizes your own strengths in a completely different subject area and appeals to just as wide an audience. In a broader context, then, there is no need to compete head-on with the competitor on every attribute, nor for every customer. It is far less expensive and beneficial for all carriers to sidestep the competition as much as meet it squarely. The first step in accomplishing this is a complete competitive analysis, as described previously.

A competitor that has technical, capacity, or coverage superiority, for example, is not met by trying to increase yours. Rather, your own superiority in advertising or control of quality distribution channels may far outweigh such advantages. Whereas you need to make reasonable advances in the competitor's area of strength, it is a mistake to try to compete on the attribute in which they have demonstrated superiority.

WHO IS THE COMPETITOR?

In the early days of wireless, competition was considered in the context of competitive solutions to wireless service, such as the public telephone, paging, and other mobile services. Once wireless technology achieved credibility as the broadest solution to mobile telecommunications access and competing systems were in place, the context of competition moved to local competition among carriers.

The development of wireless dealers and retailers as primary channels has diluted the focus of competition; often carriers consider the relationship between internal and external channels, or the promotion of service subscription versus telephone sales as the focus of the customer purchase decision, to be competitive battlegrounds. Such considerations are misplaced, and the carrier needs to concentrate on competition only to the extent that it interferes with carrier revenue growth objectives by displacing its revenue and customers; to the extent that the actions of external channels and the other carrier stimulate market growth, even the other carriers may not be true competitors.

In the future, additional new technologies may expand capacity and therefore the growth expectations of each carrier by such a degree as to make competition for customers a subordinate concern to the need to create new applications and new market segments. As a competitive threat, newer technologies and service offerings such as satellite mobile communications should realign competition at the level of rival technologies rather than carrier versus carrier.

COMPETITIVE BACKLASH

Competitive response is an important threat at all times. An important part of any campaign is to think through what your competitors' reactions will be. The more successful the campaign, the more likely the competitors will react—not just to avoid losses but because they need to save face.

While imitation is one of the sincerest forms of flattery, the best way to avoid competitive backlash is to design product features, promotions, etc., that are difficult to duplicate. Thus, a promotion to offer a new, feature-rich handset at a very attractive price might be very successful but also very easy to duplicate. If it is a unique handset, however, offered for the first time through a unique program from a certain handset manufacturer, you need only make sure that it is not offered to the competitors in your market by your manufacturing partners.

I have always been amazed, for example, that nonwireline carriers have done so little to take advantage of their freedom from regulation over their wireline competitors in each market. Whereas Bell entities were not permitted to offer information services, any nonwireline competitor could offer such services and truly differentiate their service. Today, this opportunity is lost because Bell entities no longer have this restriction.

In the same manner, however, Bell-affiliated carriers were previously not permitted to sell, give away, or otherwise participate in long-distance services. A nonwireline competitor might have simply bundled some (or all) long-distance service into a rate plan and might have a service that would not only provide great benefits to customers in long-distance savings and billing simplification, but their wireline competitor would be powerless to duplicate. As a unique promotional proposition, it would be unsurpassed. However, the window for such a blockbuster offering closed a long time ago, and now AT&T and other carriers are offering this pricing feature.

FLATTERY

If an advertising campaign is unsuccessful, not only will prospective customers be silent but so will your sales channels, associates, and personal friends. On the other hand, successful campaigns cause not only great amounts of praise from these areas but vociferous complaints from competitors and may be the best indication of the success of the campaign. One of your competitors is worried enough about its success to threaten you. Do not be intimidated. A competitive complaint about the aggressiveness of your advertising is the best compliment a competitor can pay you.

DISPARAGEMENT

Do not demean the competition. It makes wireless service look bad for all. It does not matter if one of your competitors has many more dropped calls than you do, or that retail channels do not have the depth of technical support and service—do not advertise it. Your competitor will then attack your small coverage area or some other weak point. Soon prospective customers will know every possible reason in the world not to have wireless communications, regardless of the carrier. Advertising should always portray the positive benefits of wireless service and the advantages of your own offering.

SECURITY

Many companies, large and small, do not put some of their most important strategies in writing because they are concerned about it falling into the hands of competitors (or regulators, the news media, or the general public). Business ethics dictate, first of all, that there should be nothing in your strategies that would not reflect well on the company if it were ever made public. Beyond the ethical question, it is appropriate to be very concerned about security, but it is absurd to avoid explicit, written plans about every detail of your operations because you are afraid your competitor might get them. *It is much more important that your employees know your plans than to keep them from your competitors.* Other than customer contact information and details of upcoming promotions, wireless operators and companies in general tend to overvalue information about themselves to the competition and undervalue information about themselves to their own employees.

If competitors are seriously trying to get your planning information, this is the second most sincere compliment a competitor can pay you. You are indeed a threat! Your competitors may obtain competitive information, but it is quite another thing to understand its context and act to defend against it. If you knew a competitor was developing a campaign to give away televisions, would you develop one to do the same? Probably not, unless you were already thinking along those lines. (But based on what we have learned so far, we already know that such an idea is not recommended anyway, because it does not distinguish between good and bad customers. Such a program is only worthwhile to get strong customers.) Therefore even if a competitor received confidential information, chances are slight that they would act preemptively on it.

HOW FAST MUST A SYSTEM GROW?

Many larger organizations are constrained by capacity, yet they are afraid of competition and refuse to raise prices. Early in system growth, you should be able to tell if wireless service will grow so fast in your market that you will be unable to keep up with demand. If this is true, then whichever of the systems is least congested will fill faster until service availability and customer bases are the same. If there is really a discernible difference in quality among the carriers' networks, the better will fill faster until it becomes congested. The only way to match demand to service quality in such a case is to raise prices.

Whether demand is strong or not, the sales and promotion functions should be balanced to sustain customer growth at a rate that the system can support, which does not require enormous sales and marketing resources to obtain. One look at the expected life-cycle revenue and profit per customer should tell you how much cost you want to incur to get it.

One strategy for sales aggressiveness, for example, would be to conscript every live body in the market as an exclusive dealer before your competitor does. In a growing market where there is heavy competition for distribution, this is an important and major strategy for market share. If the sales capacity outstrips the growth rate, however, this strategy leaves much to be desired: it puts too many resources after too few sales, which may eventually make the business look sour to everyone in it. It does not match sales rates to system growth and sales cost limits; it does not discriminate between professional selling and practically bribing poor-candidate customers to subscribe. In an accelerating market, it may be wise to attempt to gain every local member of a specific channel that is strategically directed at serving target market segments with this type of product and service.

A well-constructed strategic and annual plan will make sales channels, sales costs, and sales rates congruent. Each market segment provides a different amount of revenue per customer, indicating, along with customer channel preferences, which sales methods you can afford to use for each segment.

And many good external channels must be established from scratch. You cannot expect an oversupply of good candidates for wireless agents and internal salespeople in any market; many of these have to be developed. Assuming that a fixed number of quality external channel participants are available or feasible to nurture and grow, such channels may not develop for years, and the slack in the growth rate required must be made up by internal channels. If you lose the lead in developing sales channels, it is very difficult to catch up.

Those carriers that identify the best sales channel candidates early and help them succeed, will have an unstoppable long-term competitive advantage in sales.

10

THE SALES PROCESS

The discussion of the sales process presumes the training of the sales force in sales planning and presentation. All salespeople should know how to qualify prospects, maximize customer "face" time, overcome objections, and be prepared.

The retailer and company store have taken over the lead in customer volume from direct sales, which had the lead since cellular telecommunication's beginning. With increasing awareness and acceptance of wireless service, the productivity of direct sales has increased to 35 or more sales per month.

The most important aspect of the sales process for direct sales is that it must be managed, from lead generation through customer installation, as a well-oiled machine. Diligent measurement is the prime ingredient of success. Salespeople must constantly be supplied with qualified leads from outbound telemarketing efforts and advertising responses. Leads must be tracked with regard to source, assigned salesperson, time elapsed to appointment, date of sales presentation, time to installation or fulfillment, and final disposition of the lead. Within telemarketing specifically, measurement and efficiency are even more specific and more carefully controlled.

With appropriate measurement, action can be taken to discover what kinds of promotions generate the most sales, not just the most leads. The most effective people in telemarketing, sales, and customer service can be rewarded and given incentives appropriately. The relative value of telemarketing, advertising (media selection, radio format, time of day), and referrals from existing customers can be weighed and used to direct future efforts. The value of callbacks to prospects who do not buy can be determined, and the percentage of walk-in customers that results from various promotions can be recorded.

Each of these factors can be examined in the light of its success with different market segments: consumers versus business prospects, male versus female, and vertical industry classifications.

The success of the sales process is a function of management's ability to have telemarketing, advertising, and sales management function together to maximize their combined effectiveness. In addition, all of the functions must be working toward the broad goals of total sales development, rather than the specific goals of advertising awareness (advertising), lead generation (tele-marketing), and sales per salesperson (sales management). This is far more difficult than it sounds. People in each function must be measured on their individual performance within their discipline, yet they must work coopera-tively toward the larger goal. Anyone experienced in the management of the execution of the process can point to statements from telemarketing and advertising regarding the deficiencies of the sales force in completing the sale, countered by the sales department's complaints of poor leads provided.

Sales must be supported so that each salesperson spends the majority of his or her time in sales planning and the sales presentation. Yet in every organization, salespeople spend an incredible amount of time resetting appointments, completing forms, going to meetings, and tending to admin-istrative matters.

In the sales function a minimum amount of time should be spent cold-calling unqualified leads. However, lead generation is spotty in many sales organizations. In addition, support of external sales channels in lead genera-tion is often poor in both numbers and quality. External channel personnel do not have the ability to hire specialists in telemarketing and advertising, nor do they have the carrier's economies of scale for producing a quality advertising or telemarketing program. The best operations leave promotional design to a well-run carrier organization, which supplies an adequate quantity of qualified leads to all channels with field sales forces, and has the resources to facilitate a well-designed and well-executed measurement system to ensure its success.

Another weakness in the sales chain is the neglect of sharing of informa-tion before the launch of a telemarketing or advertising campaign. All involved channels, customer service, and salespeople should be briefed on the nature of the campaign, the offer, and the details about when it will appear, where to get literature and equipment, how to complete the order form and record measurement statistics, etc. Because the multifunctional nature of a major campaign, it is common that all internal communications steps are not completed before the campaign begins.

DIRECT SALES

While most of the aspects of direct sales should be covered through an extensive training program of at least five business days, there are some aspects that should be emphasized.

The direct salesperson's activities can be divided into sales planning and sales presentation. Sales planning involves the qualification and setting of the appointment of the prospect, and sales presentation involves the actual customer presentation.

SALES PLANNING

Depending on how much support in qualified leads the salesperson receives, some portion of the sales planning process includes finding new prospects. Hopefully, most of the work in sales prospecting for the salesperson involves checking back with recent new customers for referrals to qualify. Otherwise, the salespeople become telemarketers themselves most of the time.

The sales planning process can begin by reconfirming or setting the appointment and using the opportunity to find out more about the prospect. Some of the most important information includes whether the prospects are responding to a particular offer, their knowledge of wireless service and interest in specific telephones, and their business or personal applications.

For each sales visit, the salesperson should develop a strategy that includes the type of telephone to demonstrate, an order form partially completed except for optional items and incomplete customer data, the benefits and applications to discuss, and questions to ask that can help prospects understand the benefits in their personal situation. These items are prepared for each prospect as much as possible before the first sales call of the day so that the entire agenda for the day, and much of the next, can be planned.

The primary reason for failing to close a sale with an interested prospect is incomplete or conflicting information about wireless service. The salesperson needs complete information about the service, telephone equipment, prices, and competition. Probably the second most common reason is not asking for the business from a seemingly undecided prospect.

Certain aspects of the sales presentation need not require explanation: promptness, courtesy, and appropriate dress. The presentation should begin with a conversation regarding the prospect's lifestyle and business habits, and an evaluation of their understanding of wireless use and pricing.

Benefits should be discussed in the context of the customer's application and the salesperson's understanding of the business user's industry. Generally, for business users the benefits of wireless use are increased productivity and the ability to stay in touch for decisions, return phone calls and messages, and conduct business over the phone personalized to their situation. With personal wireless use the benefits that should be emphasized are avoiding second trips on errands, security in the case of road emergencies, staying in touch and time savings.

The sales presentation process can be described as a series of answering objections and test closes. If the prospect is responding to a specific offer, it can be a very brief presentation if he or she is ready to buy. Hesitant buyers can be assured that low-priced phones, prepaid billing, and economy service plans can make their decision as tentative and noncommittal as they desire.

The importance of the simplicity of the presentation and the need to answer questions honestly and to avoid dragging out the meeting once the sale has been consummated cannot be overemphasized. Optional service features, such as custom calling features and voice messaging, should only be sold if brought up in the context of the discussion.

RETAIL SALES

The retail sales operation involves almost exclusively external retailers, automobile dealers, etc., and has completely different problems from the direct sales operation. Little control can be exerted on such channels to change the way business is done; these personnel use their own advertising. Their advertising and sales efforts are rarely directed specifically at wireless services but at a broad range of electronic, communications, or other larger product lines.

The customers that approach such channels are also different. Whereas some are business users looking for a deal on a wireless phone, most are uninitiated consumers. They do not approach the wireless decision from a business productivity angle and know far less about wireless telecommunications before coming to the sales location. At the same time, the average retail salesperson's knowledge of the wireless business is far below that of the direct salesperson, whose efforts are dedicated to wireless only.

A separate sales support manager who understands retail merchandising is normally directed to help retailers manage wireless sales. The primary weakness in retail sales is the lack of information and conflicting details that uninitiated sales prospects receive about basic subjects such as service pricing and coverage and the advantages and disadvantages of analog and digital service. When we consider the turnover of retail salespeople and the number of products they sell, the job of training may seem impossible. The value of ensuring the availability of point-of-purchase literature as well as reference materials for salespeople is paramount in this environment, as is on-the-spot training for salespeople on the floor. This is easily a full-time job in itself.

The 80/20 rule applies to retail stores and other nontraditional channels as much as anywhere else. Eighty percent of sales come from twenty percent of sales outlets. Success will not come from the stores of one particular chain but from a few highly motivated well-supported locations of each retailer. This can be propagated to other locations only with a combination of the enthusiasm of the local manager and the support of the operation by the carrier.

OTHER CHANNELS

Each channel deserves a manager or managers who not only manage the effectiveness and efficiency of the channel but motivate and support its members and act as champion for the channel at the carrier organization. The internal direct sales force and external retailers, resellers, agents, and independent dealers all have different management needs, concerns, and perspectives that require separate channel managers. Retail stores need merchandising and advertising help; agents and dealers need sales training; and resellers have numerous issues different from all other channels, including billing, wholesale pricing, and service offerings, in addition to sales support. Separate managers also make channel members more comfortable that competitive information is not being leaked to other channels.

SALES PROCESS IMPROVEMENT

Wireless carriers can create a competitive advantage by organizing sales and customer services processes with which clients find it easy to do business.

Customers must feel they can spend as much time before the sale as necessary to be comfortable with their purchases. Likewise, after the sale, customers must feel that their phones are working properly and they know how to use them effectively. This implies that the salesperson must be well informed and can explain things to customers patiently in a way that they can readily understand.

The corollary to this requirement is that the customer can spend as little time as possible in the store and complete the transaction as fast as possible. Bell Atlantic has set a goal of completing activation transactions within ten minutes of the customer saying "yes" to a specific phone and service plan, exclusive of training on the use of the service. The salesperson should be able to detect subtle cues from the customer regarding how the customer feels about giving information and spending additional time in the transaction.

There are several specific areas of the sales transaction by which the customer can readily feel at ease in dealing with the company:

- Completing the activation in a short amount of time.
- Asking for the minimal amount of customer information to complete the transaction: name, address, telephone, credit information, and features requested. The amount of superfluous information often requested is unreasonable.
- Requesting demographic information and a customer satisfaction survey, only to the degree that the customer is willing.

Carriers should be as easy to do business with as they themselves want their vendors to be.

11

CELLULAR SALES PRODUCTIVITY[1]

Sales productivity is a major concern of all wireless business people. Whereas sales managers work hard to increase the number of sales per salesperson, there are other ways to analyze and improve sales productivity across all sales channels. One method is to analyze the productivity of each sales channel and optimize a mix of several sales channels based on their efficiency and appropriateness for specific kinds of customers.

Manufacturing businesses can associate current sales revenue with current sales and marketing costs, because most of the revenue due to the sale of a product is immediately reflected in the income statement. This is much more difficult to do with a subscriber service business such as wireless service, but it is necessary to understand the relationship between sales costs and revenue in order to determine if its sales cost is an efficient generator of net income. We will use a wireless carrier here to illustrate how such an analysis is done, but resellers, agents, and other wireless business people can use a similar method.

The best way to measure the efficiency of the sales effort of a subscriber service is to compare the current sales cost to the revenue that will result from the sale over the time that the average customer keeps it. This most closely relates the sales effort to the revenue it generates, in a way similar to the sale of a manufactured product. An average customer life of 30 months equates to an attrition rate of $3\frac{1}{3}\%$ of the customer base per month. If an

[1]Adapted from the article "Cellular Sales Productivity," which originally appeared in the April, 1992 issue of *Cellular Business*. Copyright 1992, Intertec Publishing Company, Overland Park, Kansas.

average customer bill is $80 a month, then the cost of each sale in the current month will yield 30 × 80 or $2,400 at retail in the future. We will call this the customer life revenue to distinguish it from current revenue on the income statement.

The wireless carrier must determine the cost of selling through each sales channel (direct sales, agent, dealer, reseller, etc.) and compare it to the profit that will result from the $2,400 in revenue it generates. A typical income statement for a wireless carrier is given in Table 11.1, based on percent of revenue.

While it is important to keep all categories of expense under control, the sales expense of this period does not generate this month' revenue but *future revenue*. Minimizing sales and marketing expense as a percentage of current revenue saves cost but also reduces the ability to generate future revenue.

The income statement in Table 11.1 shows that any revenue, current or future, that Alpha Cellular generates produces only 40% of revenue in operating income before taking into account the discounting of future revenue streams. In our example, $2,400 of customer life revenue will produce only 40%, or $960, in future operating income. If the carrier really wants to achieve an operating income of 40% of revenue within this cost structure, its long-run sales and/or marketing expense must remain at or below 13% of revenue, as shown in the income statement. To accomplish this, the sales cost must also not exceed 13% of the customer life revenue of $2,400, or $312. How many cellular carriers can say they are really doing this?

Unfortunately, many operators are so driven by subscriber numbers that they do not afford themselves this kind of rational analysis. The sales team is directed to get subscribers of any kind at almost any cost, and the financial manager and general manager wring their hands trying to cut sales and other costs indiscriminately.

One of the first steps toward such a rational analysis is to attempt to construct an income statement for each sales channel. Hopefully, there are

TABLE 11.1   Alpha Cellular Company Income Statement as Percentage of Revenue

Revenue		100%
Expenses		60%
Sales and marketing	13%	
Administration	7%	
Operations	20%	
Depreciation	20%	
Gross operating income		40%

records that show which sales channel has sold each customer and the current revenue for the set of customers attributed to each channel. Marketing costs such as advertising, which may be associated with multiple sales channels, may be allocated based on the proportional revenue or direct cost of each channel to which it applies. It is a real eye-opener to see the revenue and cost of each channel detailed separately.

The easiest division of channels is to separate the reseller channel from the aggregate retail channels (direct sales, agent, dealer). The reseller channel may produce lower revenues at a lower wholesale price, but its costs should be lower also. It is possible that the reseller channel may be more profitable than the retail channel. Not only are retail sales commissions eliminated when sales are made through the reseller channel, by the customer support function, billing function, and collection or bad debt costs are all lower. For comparison purposes, only those costs that differ among channels need be represented. Thus operations, administration, maintenance, and depreciation expense per customer, for example, might be assumed to be the same in all channels. An illustrative version of such statements is shown in Table 11.2.

The total channel cost can be divided by the gross sales gain to determine the cost per new subscriber by channel. However, not all customers are of equal worth. The total customer base for each channel should be divided into its revenue stream to derive current revenue per customer by channel to determine the quality of customer each channel is producing in terms of

TABLE 11.2 Alpha Wireless Company Monthly Income Statement by Channel

Reseller Channel			Sales Summary	
Revenues		$128,000	Gross gain	40
Expenses		4,700	Customers	2,000
Marketing	$2,200		Revenue per customer	$64
Customer service	1,200		Cost per gain	$118
Billing or collections	200		Customer life revenue	$1,920
Sales	1,100		Revenue/cost ratio	16
Revenue less channel expenses		$123,300		

Retail Channel			Sales Summary	
Revenues		$1,120,000	Gross gain	400
Expenses		224,000	Customers	14,000
Marketing	$57,000		Revenue per customer	$80
Customer service	15,000		Cost per gain	$560
Billing	32,000		Customer life revenue	$2,400
Sales	120,000		Revenue/cost ratio	4
Revenue less channel expenses		$896,000		

revenue per month. This is then multiplied by the average customer life to determine the customer life revenue. The customer life revenue, remember, is not the same as income statement revenue shown here; it is merely a measure of channel efficiency when compared to the cost of the channel per sale.

The reseller average revenue per customer is $128,000/2,000 or $64 per month. The customer life revenue is 64×30 or $1,920. For the retail channel, the average revenue per customer is 80×30 or $2,400.

In the illustrative monthly statements shown here for the composite of retail channels versus the wholesale or reseller channel, the retail channel produces far more revenue and customers than the reseller channel. However, the cost per new subscriber is much higher than that in the reseller channel. If these results were not separated by channel, the combined results would show a gross operating income of about 38% of revenue, close to our composite example. In reality, however, the retail channel is operating at a level of only about 33% profit, whereas the reseller channel is at 50%.

The most important number here is the revenue to cost ratio. The ratio of customer life revenue to sales cost for the reseller channel is 16, whereas it is only 4 for the retail channel. The reseller channel is much more efficient than the retail channel.

How would you feel, as general manager, if you knew that it took $560 in sales costs to generate a gross income over 30 months of only $960? You may find that out if, as the manager of retail channels, your results are like the retail illustration shown here. Instead of gaining customers at any cost, you might determine that unless low-usage consumers were willing to buy directly over the phone in response to an advertisement, you may not want to attract them with free wireless units. You may consider structuring commissions and bonuses based on the market segment of the customer.

This invites the same type of analysis of each individual retail channel—direct, agents, dealers, retailers, etc.—to the degree that the information is reasonably easy to obtain. This would show which of the retail channels is the high-cost culprit or if they all are. Those channels producing the highest-revenue customers would also be revealed. Channels producing customers at good revenue to cost ratios should be expanded, but only if market research shows that there is significant potential left in that market segment. It must also be considered that a channel cannot be expanded if there is little talent available with which to expand it. High-cost channels that are also not concentrating on high-revenue customers will be revealed by a low revenue to cost ratio. In contrast, expensive channels may not necessarily be the least efficient if they are producing high-revenue customers.

The direct sales force is probably the most expensive sales channel. It should be matched to corporate customers who need the kind of support this channel can give and whose usage characteristics can support those costs with high revenue. Since this channel is normally associated with premium corporate accounts, it is often given the most marketing attention, the highest-quality sales leads, the most professional sales resources, and other expensive support that is difficult to assess objectively without careful analysis. However, this channel may not only pull in customers with higher monthly usage; they may also be more loyal and stay with the service longer to produce more revenue.

If a different customer life can be calculated for each channel and used in the customer life revenue calculation, it should be done to show the truest picture of the quality of each channel's customers. A newer channel may produce distorted results in longer customer lives because little customer attrition has set in yet. If a channel is selling to a special market segment, it may be a characteristic of the segment that causes greater loyalty, rather than a reflection on the quality of the sales effort of the channel currently selling to it. So it is also important to measure revenue and customer life by market segment, regardless of the channel selling to it.

Agents and retail dealers often bring in the bulk of new customers from the largest market segments. Commission levels may make the channel very expensive in a competitive environment. The opportunity to bring in retail storefront channels (or other channel alternatives) at lower commission levels for low-usage consumer segment customers is an option and may be an attractive channel to these consumers. However, it must be supported by a marketing program that matches the needs and costs of this segment and channel.

As wireless service grows, it must continue to attract more users from diverse market segments. In the larger view, the current business environment is experiencing a fragmentation of traditional market segments into smaller and smaller subunits—even 'niches'. Each of these subsegments must be attracted to wireless service according to their needs and applications, but also with attention to their channel preference and at a channel cost matched to their contribution to the business. Sales and marketing managers must become more attuned to the financial realities of marketing rather than attract any type of wireless customer at any cost.

As wireless subscriber numbers expand, more and more marginal users are attracted, who generate less than average revenues per subscriber. Therefore the carrier must begin to understand which market segments generate the most revenue and direct their highest-cost sales channels to those potential users. Other segments that generate less revenue must be attracted to lower-cost sales channels.

If we understand the revenue per customer for each market segment, the cost per sale of each channel, and the channel preference of each segment, we can assemble a matrix that matches channels to segments and exposes gaps in the approach, such as the one shown in Table 11.3 for Alpha Wireless Company. Only a few significant segments are shown here for illustration; the small-business segment has not been included. The monthly revenue is multiplied by 30 months (oversimplifying the calculation by assuming the same average life) and divided by the cost per sale to calculate the customer life revenue to sales cost ratio, referred to in the table as "Ratio."

This illustrative analysis for Alpha Wireless Company shows that the direct sales force costs too much to support the current corporate or large-business segment because the revenue is not high enough to support the cost. Some of this business should be switched to agents unless another alternative is proposed. The consumer market has an even lower revenue per customer and desperately needs a lower-cost "retailer" type channel, which research might show the customer would prefer. The reseller channel cost is so low that it should be expanded to attract other segments that are not attracted to retail channels or hard to reach. It may be a candidate channel for the construction segment, the revenue of which does not support the sales cost in this market. A radio dealer who specializes in construction could become a successful wireless reseller with his or her existing customer base.

These are just a few of the actions that might be suggested by the analysis for a particular operation. Other wireless business people can analyze a similar breakdown of their own channels and segments. Agents, for example, might want to do the same analysis to compare their direct sales, telemarketing, and dealer support channels.

TABLE 11.3   Channel Analysis for Alpha Wireless Company for the Market Segment

		Business			
	Corporate	Professional Services	Real Estate	Construction	Consumer
Monthly revenue per customer	$200	$385	$270	$140	$60
Current channel	Direct	Agent	Package reseller	Two-way dealer	Agent
Current base	2,000	400	325	210	460
Potential	35,000	3,500	1,200	625	140,000
Cost per sale	$740	$560	$110	$450	$560
Ratio	8	21	74	9	3
Gap	High cost			Low revenue	High cost and low revenue
Preferred channel	Direct	Agent	Reseller	?	Retailer
Recommended channel	Agent	O.K.	Expand	Drop	Low-commission retailer

The process of producing and analyzing this data is as valuable as the result. Just realizing the need to see how sales channel effectiveness, sales channel cost, and segment revenue interact will produce large benefits in changes that increase profitability as well as sales levels. With such changes one must carefully consider customer needs and other factors, such as the ability of a channel to expand.

Traditionally, marketing and sales channels have been asked to meet revenue and customer gain goals under budget control. More and more, they are being asked to relegate their efforts directly to productivity and profit. This becomes more necessary as markets fragment and margins shrink. Such an approach will not only improve profitability but also improve the quality of the customer relationship and the effectiveness of the sales channels while introducing better communications between the marketing and financial staff.

12

THE LIMITS TO GROWTH[1]

Sales managers are under constant pressure to achieve subscriber gains. It is unfortunate that the sales function is given the responsibility for net gain, since it has little control over most customer losses. While the sales channel can do more to make new sales "stick" and to keep existing customers satisfied, much of the problem is concerned with uncontrollable churn. However, this chapter is not about churn but about keeping growth on track.

While sales managers are used to being hounded for better sales results, their bosses do not always understand that a large part of the problem in increasing the customer base has nothing to do with sales at all. Sales managers may want to tear out this chapter and show it to their general manager, if only to vent their frustration.

When a wireless market first starts up, almost none of the customers leave the service until about 18 months after service begins. Churn is not even noticed until about three years have elapsed; a monthly attrition rate of 2% to 4% becomes a substantial number when applied to a customer base of reasonable size.

If we do not recognize this as a phenomonen from the beginning, it may appear that sales productivity is slowing down, when in fact it is probably improving. Thus we need to separate net gain into its component parts: first, the gross gains in customers due to the sales effort, and second, the natural

[1]Adapted from the article " Limits to Growth," which originally appears in the Myths, Tips, & Facts supplement to the July 1992 issue of *Cellular Business*. Copyright 1992, Intertec Publishing Company, Overland Park, Kansas.

loss of older customers from the base due to customers changing jobs, moving, or becoming dissatisfied and moving to another carrier.

If the customer base continues to grow, but the sales effort stays constant, evenentually net gain will be sharply reduced through no fault of the sales effort. This is the basic premise of what I call the *Steuernagel syndrome*:

> Given any fixed amount of sales effort and any churn rate, net customer gain will virtually reach zero in a few years.

A simple model that anyone can create in a few minutes with a spreadsheet will demonstrate this phenomonen. Just enter a constant sales rate of, say, 100 gross additions per month. Each month, add these to a total. Then apply an attrition rate to the total, for example, $3\frac{1}{3}$%, and subtract it from the total as churn. Then repeat the process over many months.

Using these numbers, Table 12.1 shows the results. We can see that after only 12 months, our net gain is only 70, or about $\frac{2}{3}$ of our sales rate of 100. After 24 months, we realize a net gain of only 46% of our additions; after 36 months, a fixed sales capacity of 100 per month produces a net gain of only 29 customers and a growth rate of just 1%. While the sales force can contribute to the effort to lower churn through customer support, most of the losses are probably not due to customer dissatisfaction.

Once we realize that we can only count on a customer for a certain term (a $3\frac{1}{3}$% monthly churn rate is equivalent to a 30-month 'customer life'), we can recognize this in our financial and sales planning and address the problem by applying the corollary to this Steuernagel syndrome:

> Sales capacity must continually expand to achieve growth in the customer base.

TABLE 12.1   Model of the Steuernagel Syndrome

Month	Additions	Losses	In Service	Net Gain	Growth Rate
1	100	0	100	100	100%
2	100	3	197	97	97%
3	100	6	291	94	48%
⋮					
12	100	30	1,009	70	7%
24	100	54	1,684	46	3%
35	100	68	2,104	32	2%
36	100	69	2,135	31	1%

We can calculate how rapidly the sales rate needs to climb in order to maintain a constant growth rate, but we must also consider the ollowing parameters in expanding sales capacity:

1. Sales productivity, or the number of sales per sales cost dollar expended, must remain constant or increase;
2. Existing sales channels cannot be expanded indefinitely; new sales channels must be found and exploited;
3. Target market segments and geographic territories must be broken down more narrowly among the sales channels, and new target markets uncovered;
4. New applications, service features, pricing plans, and sales opportunities must be developed to match new channels and customer segments.

INCREASING THE SALES CAPACITY

Month 36 of service in Table 12.1 was the first month in which the growth rate of the customer base dipped to 1%. In order to keep the growth rate at, say, 2% per month, or 24% per year, we merely "size" the sales capacity required and increase it accordingly instead of keeping the sales capacity constant. Using our example again, instead of maintaining the sales capacity or rate at 100 per month, starting in month 36 we calculate the sales rate required to be equal to the attrition, or churn rate, plus the growth rate desired, times the customer base for month 35:

$$\text{Growth rate} + \text{Churn rate} \times \text{Customer base} = \text{Sales capacity}$$
$$2\% \quad + \quad 3.3\% \quad \times \quad 2{,}104 \quad = \quad 112$$

So in month 36 we must increase our sales capacity from 100 to 112 to keep our customer base growth rate at 2%. Once we make this adjustment, all we have to do is keep increasing the sales force at the same rate as the desired growth rate—but we must constantly increase the sales capacity thereafter, or the growth rate will drop to zero. In our example, this means that we will have to more than double our sales force every three years.

OTHER USES

Some of the earliest wireless markets have arrived at these conclusions through trial and error. As the sales capacity is expanded, the existing salespeople, agents, and resellers may express concern that the expansion of the sales force is unnecessary and competes with their efforts. The need to diversify the sales effort and increase it at a management-controlled rate requires the carrier to add sales channels and channel members as well as enlarge existing ones. The demonstration shown here can help allay the fears of in-place sales channel members through providing an understanding of the phenomenon.

If we look at an agent instead of a carrier, we can see the same flattening of growth through this analysis. If a new, conventional wireless agent starts up with a sales force of 20 members each of whom produce 25 sales per month, its nominal sales rate will be about 500 per month. Yet this analysis will show that unless productivity increases, the sales force is enlarged, or the agent adds other sales methods, the agent will "top out" in three to four years at 10,000 to 12,000 customers.

The flattening of the growth rate for an agent does not mean that performance is at fault. A new agent might look attractive compared to an existing agent because churn has not yet become a factor, but in reality the new agent is not performing any better than the in-place channel. The solution is to include this analysis in the forecasting of gains over the next three to five years and to seek additional channel members and new channels that meet the needs of emerging market segments constantly while supporting existing channels in productivity improvements and new sales methods.

This analysis can also be used as the basis of an analysis to understand the reasons for rapidly declining customer base growth rates as a wireless system matures, as a sensitivity analysis for the effects of increased and decreased churn, and to demonstrate the portion of increased sales capacity that can be met through productivity increases.

Most intriguing are the implications for the sales force size if an analog cellular carrier in a top 10 market expands its nominal service capacity through digital technology by a factor of 10 or more. Our model may infer that finding sales resources for filling digital wireless systems may be even more difficult that deciding on the technology standard or finding the customers, and that it may be unfeasible for competing carriers to find the sales resources to fill a vastly expanded total system capacity in a reasonable time frame.

13

PRODUCT MANAGEMENT AND DEVELOPMENT

PRODUCT MANAGEMENT

The typical wireless carrier will have most of the following additional services available, some of which we have already mentioned:

Custom calling features: call waiting, call forwarding, three-way calling, and conference calling

Voice messaging

Handset insurance

Voice dialing

Long distance

Information services: traffic, enhanced directory assistance, roadside assistance

Wireless data (more appropriately, a platform or application rather than a service)

Digital features: messaging, paging, caller identification

Other services, like prepaid calling, may be considered products or special pricing plans.

These services require a separate product management effort, in addition to the product management effort for general wireless voice service. The actual product management effort may be combined into a single position.

Each of these services should have an objective, strategy, and forecast. The objective may be to generate profit, provide additional customer convenience, add to the breadth of the offering to customers, or generate usage. The strategy should aid this objective.

Therefore some services might be free or packaged with air time in order to generate usage, whereas others may not be highly subscribed but are excellent discriminating competitive features for the carrier in the marketplace. The positioning and advertising of each service in relation to each other and the main wireless voice product should be clear and understood by everyone and should further its particular objective and strategy.

Products and product management can be organised in several ways. Products may be grouped according to market segmentation, but the latter is usually used as one input from market management into product management. Another way is to group products by objective: profit, ease of use, or usage stimulation. A better way is to organize products by the way they are implemented and presented to the customer. Products are often categorized as follows:

1. *Pricing Features*: features that can be implemented merely by introducing pricing changes.
2. *Service Improvements*: products that are available to all users without cost or enrollment.
3. *Enhanced Services*: products that are available to users as an option at a price.
4. *New Services*: services that can be used and are valued by customers, independently of wireless voice services.
5. *User Terminal Equipment*: handsets and other user terminals, accessories, and associated services such as warranties and repairs.

PRODUCT CATEGORY EXAMPLES

Pricing Features: calling plans, prepaid calling, home zone pricing, home rate roaming, first incoming minute free.

Service Improvements: coverage, capacity, automatic call delivery.

Enhanced Services: enhanced directory assistance, voice dialing, voice messaging, information services.

New Services: wireless data, dispatch, wireless local loop.

User Terminal Equipment: handsets, data terminals, modems, batteries, chargers, wireless fax machines, handset insurance, etc.

Each service should be tracked by the product manager according to how well it meets its objectives. Those with profit objectives should have their own profit and loss statement, showing how costs and revenues are meeting objectives. In addition, new features and products should be tracked against the business case results used to evaluate the product for commercial offering. This is rarely done and is important to test whether the product is meeting the financial and other objectives postulated for it in the planning stages.

Other, more complicated performance measures may be required for products the objective of which is to increase usage or help to attract new subscribers. Churn, average life, and other parameters of features can be valuable in evaluating feature performance. The role of the product manager is to take steps to achieve the feature's objectives within budget.

The product manager may have pricing responsibility, but this function may be assumed by the pricing manager or marketing director with input from the product manager, especially if the product has a more strategic rather than volume or profit objective. The product manager must make sure that the product meets customer needs and is periodically enhanced to do so; to make sure all product communications—literature, advertising, descriptions, press releases, etc.—reflect the product correctly and in the best light; and to make sure that the product receives the promotional, sales and advertising attention it requires to achieve its objectives. He or she must make sure that the product has proper positioning within the sales, sales incentives, advertising, literature, promotional, and other programs in which the product may have only a small part. This requires intensive work with other product managers and marketing communications personnel as well as with the sales organization.

NEW PRODUCT DEVELOPMENT

Features such as voice messaging and information services have been developed for the wireless industry in parallel with their development in the wireline industry. New product development has not been successful for wireless carriers and will probably not be in the future. Instead, carriers will need to develop new telecommunications businesses to complement their current mobile phone business.

New product development in cellular and PCS has failed for several reasons. First, product development requires an expensive research and development effort that few carriers can support. Those that have a large research and development staff have found more success in making engineering improvements to their networks to improve efficiency and lower costs. It is difficult for carriers to make product innovation pay off in their own markets, and few have gone into the business of selling their innovations to others. Therefore it is more economical for a carrier to have its infrastructure vendors and other vendors develop new products and services.

Vendors have not been able to deliver a timely product to carriers, however. The products delivered are homogenized versions of product requirements given the vendor by multiple carrier customers that may not meet the needs of any individual carrier. Yet a custom product from a vendor is almost as expensive as it is to develop yourself. Vendors also need long lead times to develop new products, and the market window for a new product may be long past by the introduction date.

Too often, carriers will implement a new product or feature in all markets by purchasing a large amount of system hardware only to find it has low demand. A better product process for a carrier is to identify a product either through its own experience or through customer needs assessment, to product a successful business case, and then to create a substitute with existing technology in a test market. The market attractiveness can be tested in this way. With customer demand, interest from other carriers and manufacturers in the product can be raised to make it less expensive to develop. The important part is to be able to test the product with minimum risk before paying for a major development effort and implementing it at major expense.

Wireless data service, for example, has been slow to grow as a product. Much money could have been saved if the demand could have been tested using one of the existing narrow-band radio data services. Several wireless carriers have been supporting the development of cellular digital packet data (CDPD) since 1992 and have deployed it in their networks. Yet United Parcel Service, the largest user of wireless data in the United States, has consistently used regular circuit-switched cellular voice channels for its data application.

Wireless data terminals for mobile executives and vertical applications alike have been mostly unsuccessful, yet many carriers have helped fund terminal development for such applications. Whereas such terminals are necessary for the success of wireless data, the existence of the world's greatest wireless terminal will not sell wireless data. Also, investing in the development of such a terminal will never pay off for a carrier that can only sell it in its own markets. Customer need will determine the timing of the success of wireless data, and carriers need not take the lead in developing solutions. A demand for the use of voice channels for data applications will be a signal that a more efficient means of data transmission is required. Meanwhile, competition has forced down the price of wireless voice channels to make it more competitive for low-speed data.

Although carrier product development is not favored, several assumptions and observations must be made regarding wireless product development in the U.S. market:

1. Intelligence, functionality, and low cost will always gravitate toward the terminal, not the network.
2. Carriers cannot make enough money on a new product or feature to justify a basic research and development effort.
3. As a general rule for product development, carriers function as distributors of new vendor technology to end users and are losing money if they do otherwise.
4. Nationwide and industry-wide standardization requires a vendor-developed product; other carriers will not adopt a proprietary technology invented by a carrier as a *de facto* standard.

The role of product development in the product management organization of a wireless carrier is to:

1. Evaluate user needs and new vendor technologies and products
2. Prepare business cases for new feature and product commercialization
3. Specify products to the engineering department and to vendors to meet company and customer requirements
4. Act as project manager for new products from concept to implementation as a market-driven strategy

New product development is guided by customer input, but this cannot be the only input to the process. New applications developed by manufacturers of infrastructure equipment as well as other vendors and the success of other carriers with new features and applications are also indicators of what should be offered.

The carrier must have a clear idea of the types of new services in which it is interested—those that produce income, that differentiate the carrier competitively, that simplify use of the service, that stimulate usage, or that position the company as an industry leader. Some enhancements, such as coverage and capacity increases, may not fit discrete product areas or produce revenue but are just as important as new features.

The wireless carrier, however, is not a basic research organization. It cannot benefit enough from basic research to fund it, as it has no capacity to leverage any new technologies into licensing, manufacturing, developing, or selling such products to markets other than potential subscribers. In a sense, we can view carriers as distributors of the products and services of infrastructure, handset, and other vendors to end users.

Product development for the wireless carrier is therefore a process of evaluating customer needs, offerings of competitors and of other similar carriers, and the product platforms available from vendors in order to come up with those products, services, and features to offer to its customers. This must be done in the context of the strategic objectives set out for current and future products. Carriers must decide whether they are industry leaders, product innovators, premium service providers, or price leaders and must determine their objectives and strategies for product introduction within the context of these objectives and those already set for the various product categories.

Some method of categorization of new products similar to that presented previously is recommended. This permits each product manager to assess new products in the same category only. This categorization also helps determine the vendor products that need to be evaluated most closely by engineering and other technical personnel to find the best way to implement them. Otherwise the product managers and technical support personnel might have to examine all proposed vendor products. Alternatively, a separate product development area can evaluate new products.

THE PRODUCT DEVELOPMENT PROCESS

The development of new products requires some discipline to ensure that it is done objectively. Unfortunately, the formalization of the process usually becomes cumbersome, lengthy, and costly in most telecommunications organizations. Few products therefore are actually developed; those that are actually commercialized were successful in sidestepping parts of the development process rather than completing them.

Without such a process, products are developed at the whim of executives, by the power of the engineering department to purchase new technology, and by the sales acumen of vendors. The carrier is left with a hodgepodge of unsuccessful and unrelated products, which additionally, are often never discontinued if unsuccessful. A simple yet disciplined method of evaluating products is required that ensures that only products with a reasonable probability of success are implemented.

Such a process may be divided into three phases, which can be completed in 90 to 180 days: (1) definition, (2) feasibility and design, and (3) business case analysis.

The definition phase describes the customer need and other secondary inputs to demonstrate that the product will provide substantial revenue, solve customer problems, provide competitive differentiation and leadership, improve the current offering, and otherwise substantially achieve its stated objectives. In this phase the product is also described regarding user functionality and distinguishing features in detail in a way that demonstrates the ability of the product to satisfy objectives.

In the feasibility and design phase, engineering, billing, and other technical contributors join product management to determine the best way the product can be designed to achieve the product definition at a reasonable cost.

In the business case phase, a complete financial analysis is completed, with a detailed forecast of demand, revenue, revenue stimulation, costs, cost savings, and other financial parameters. The final business case analysis includes a complete technical description of the product, its marketing and implementation program, and justification of the forecast and intended market segments. The analysis should demonstrate that the product can achieve some benchmark measure of profitability or other strategic objectives.

Following a successful business case analysis, it is important to use a market trial in a restricted geographic area to test the demand and performance of the product. If possible a prototype, a close substitute, or vendor-provided equipment should be used to minimize the economic risk. If successful, commercial deployment of the product should commence in stages through multiple markets or segments. After successful commercialization, the product would be turned over to the appropriate product manager.

An effective organization acknowledges failure of a product in trial or early deployment if indicated by lack of demand or poor performance as soon as possible and without unwarranted blame on those responsible for its development. Management approves the product, not the product developers, and takes the blame for product failures. Otherwise, no new products will be proposed.

14

CUSTOMER CHURN AND LOYALTY PROGRAMS

Churn is the telecommunications industry's term for customer attrition, taken from the more general meaning of the effect of different customers subscribing and canceling service without any net gain.

As we have discussed earlier the Steuernagel syndrome in Chapter 12, churn is of little concern to the carrier until the size of the customer base increases to the point of making the absolute size of churn a matter of concern. Then replacing churn becomes a major portion of the gross sales requirement, and channels must grow even more rapidly. Today, for wireless carriers, churn hovers around 25% to 35% per year, while wireline carriers have churn rates of only 10% to 20%. Because many LEC's still have a relative monopoly for local service (an observation with which they might argue), we can assume that most of the additional churn for wireless carriers is due to competition and the discretionary nature of wireless service versus basic local wireline service. This means that wireless churn can theoretically be controlled down to about 15% per year or about 1.25% per month.

The cost of churn is enormous. If we assume the acquisition cost of a new customer is $300, then a churn rate of 2% for a base of one million subscribers costs $72,000,000 annually to replace. From another perspective, if these same 240,000 churning customers could be kept 6 months longer at the $42.78 average subscriber bill for 1997 (according to CTIA), revenue would be $61,603,200 higher. These two effects can be combined, but reducing churn moves cost savings directly to the bottom line, whereas revenue has costs and taxes to be taken into account. A carrier with a million subscribers can raise its valuation by $100 million if it decreased churn by 20%, or about 5 percentage points per year.

Churn has much broader implications for growth, profitability, market share, and market leadership. Some may argue that it is too broad to be a marketing or sales issue. We have already stated that sales performance should be based on gross rather than net additions, because the sales department cannot be held responsible for customer churn. If the sales channel is selling to the right target market, educating customers properly, and putting them on the right service plan, the sales personnel are meeting their responsibility for churn prevention.

Churn is an important concern in marketing for several reasons. First, churn problems should motivate marketing to attract longer-lived, higher-usage customers to sales situations, to increase profitability. Second, marketing is responsible for loyalty programs, customer service performance (in some carrier organizations), customer satisfaction, and customer communications, with all of these functions directed at keeping rather than attracting customers.

CHURN EVOLUTION

Churn has undergone an evolution in its causes and treatment. Controllable churn was first caused by wireless service being too expensive. With the advent of a second carrier, and more recently additional carriers in the market, switching to another carrier became a churn issue, more because of coverage and quality issues than pricing.

As markets began to accelerate in growth, unscrupulous independent dealers would purposely churn customers among carriers—deactivating them from one carrier and subscribing them to another to receive an additional commission. When carriers added clauses to dealer contracts to require that customers stay on the service a minimum of 90 to 120 days to receive a commission, some dealers would churn customers when the minimum service period to receive a commission was completed.

Churn has increased in importance in recent years due to several factors. First, with the advent of PCS carriers, competitive pricing and promotions have become more of an issue. PCS carriers have been more competitive in pricing because of their need to attract potential competition from existing cellular carriers, as well as to attract new, more price-sensitive, marginal segments. PCS carriers have also introduced new services to help make their offering more distinct, and the desire to switch to a PCS or alternative cellular carrier to upgrade to digital service is also a factor. Some PCS carriers have removed the service contract requirement. These actions and others have caused cellular carriers to be just as competitive in response. Serious customers found that some carriers with low prices and innovative services had poor coverage and switched back to a more mature carrier. With all the competitive and aggressive new offers continually coming into the marketplace to gain customer attention, wireless customers can experience anxiety that they are seriously overpaying for service.

Switching carriers has thus become an increased source of churn. It is expected that average churn levels for wireless carriers may rise to 40% or more. Several other factors have increased management attention to churn. Sales and commission costs are rising with carrier competition for major retailers as the most attractive channel for the bulk of new personal-use customers. These customers have reduced ARPU because of a more discretionary need for the service. This means that they can easily switch or cancel service without as much concern for loss of use or changing numbers as a more traditional business customer, who finds the service an absolute necessity.

High sales costs, lower ARPU, and shorter customer life mean lower profitability per customer. This directs carriers to attract higher-profit customers—those with longer lives, higher usage, and greater need. However, such customers are hard to find, and churn losses urge carriers to find any type of customer to produce net gains in customer count and total revenue. Thus carriers cannot solve the profit squeeze alone by being more selective in gaining customers.

Table 14.1 gives only a general classification of controllable and uncontrollable churn. One could say that the bill is not paid because of one of the controllable causes, that carriers can make arrangements for

TABLE 14.1   Causes of Churn

Uncontrollable	Controllable
1. Moving	1. Price
2. Loss or change of job	2. System problems: coverage, dropped or blocked calls
3. Bill not paid	3. Competitive offer
	4. Poor customer service
	5. Wrong service plan for changing needs
	6. Fraud problems
	7. Lack of phone upgrade or upgrade to digital phone on competitor

service for a customer with another carrier when they move, and that a change in service plan could save a customer who lost or changed their job. Likewise, some aspects of controllable churn could be deemed uncontrollable.

The result of the acceleration of churn and the other factors decreasing carrier profitability is that carriers are recognizing that *customers cost five times as much to get as to keep* and are looking aggressively for ways to prevent customers from leaving as much as for ways to attract new ones. The unanswered question is, if you keep an existing customer, is his or her service extension only one-fifth the life of a new customer?

The ways to keep customers involve short-term methods of predicting who will churn and preventing them from doing so and long-term customer loyalty programs that make it difficult for customers to leave (or easy for them to stay). Predicting and preventing short-term churn includes (1) the use of software to analyze symptoms leading to churn and (2) the use of a special customer service retention team to handle customers who call in to deactivate service and to call such customers proactively and prevent them from leaving.

Churn Prediction

Churn prediction software can check for symptoms that indicate that a customer may leave the service soon. These symptoms can be tracked through good customer record databases and billing history and include:

- Changes in usage habits—higher or lower usage that may make a customer's current service plan expensive
- Increasing frequency of calls to customer service
- Increase in the elapsed time between invoice and bill payment and increased late payments
- Service contract near expiration

Such customers can be called and offered various incentives to stay on the service, depending on their usage and longevity.

Customer Retention Unit

A special unit of customer service can be assigned to combat near-term churn. Members would be specially trained in understanding how customers may be subtle in indicating their intention and reason for churning, offering incentives and solutions to save the customer. Customers calling in to customer care to deactivate are switched to these specialists. A proactive group calls customers identified as imminent churn candidates by churn prediction software applications. These units report a success rate of 40% or more in keeping customers from deactivating and offer such incentives as air-time credit or a phone upgrade, depending on the type of customer, usage, or longevity.

Loyalty Programs

Another measure to reduce churn is loyalty programs. Various methods can be employed to keep customers loyal by providing continuing incentives to stay on the service. In the standard loyalty program, customers are awarded points based on their usage level, which they can accumulate toward rewards in gifts, travel, or merchandise.

While some programs provide airtime and accessories, many business users can claim these items as expenses and find gifts more valuable. Carriers find that users do tend to stay with the service longer in order to receive these incentives. The most valuable programs are tied into well-known national programs for airline mileage. Some carriers try to put their own mark on the program by providing their own brands of gifts and points and feel that they can reduce the cost of the program by doing so. This devalues the program to the customer.

The most valuable long-term programs for customer loyalty are (1) the general devotion to quality service that makes an individual customer service experience so memorable to a customer that he or she is loyal for years afterward and (2) features and services that are so useful that the customer feels that he or she cannot conveniently leave the service. Customized billing for corporate customers is an example of a feature that is peripheral to service yet is difficult for the customer to give up and redesign with another carrier.

While carriers pay lip service to customer service, few have lived up to customer expectations and benchmarks for answering the phone immediately and resolving problems with one phone call. The carriers that do help promptly should explicitly remind callers that they answered immediately and were able to provide a solution at the end of the call. Carriers that keep customers waiting while reminding the caller that "your call is important to us" are only inviting customer hostility.

Customer Service Processes

A set of customer service practices will help to ensure customer loyalty and provide a competitive advantage:

- The customer will reach a live customer service representative within 30 seconds of call completion.
- All customer service requests can be handled by phone if the customer desires, for example, sending a problem phone to the carrier instead of bringing it to a company location.
- The carrier is so confident that the phone has been answered promptly that the customer service representative asks, "Were you kept waiting long?", to reinforce to the customer how rapidly the phone was answered.
- The customer service representative is empowered to provide credits, air time, or other remedies to most customer requests without consultation with a supervisor or callback in 90% of cases.
- The customer can resolve their customer service problem with one phone call—"one-call resolution"—because of the policy mentioned previously.
- The carrier provides the customer with a free detailed bill with explanations of charges.
- A schedule of calls to the customer at regular, well-spaced intervals to make sure the customer is satisfied, has the right rate plan, etc. is devised.

The above schedule of calls may include:

1. *Welcome Call*: protects the carrier by making sure that subscription fraud has not taken place but also ensures that the customer is satisfied, is on the correct rate plan, and knows how to use the phone.
2. *At 30 to 90 days*: to see if the customer has any need for additional features or a different service plan once he or she has established a real usage pattern; to see if there is "sticker shock" with the first bill.
3. *At 180 Days*: to determine satisfaction and to see if advanced features are necessary.
4. *Before contract expiration*: to determine if the customer is vulnerable to cancellation, to make a special offer for retention, and to determine if a phone upgrade is necessary.

15

A NEW ERA

In recent years, several developments in the industry indicate vast changes required for carriers to stay competitive:

- The entry of PCS carriers has reduced the general price level and forced incumbent carriers to change prices and policies, for example, offering service without contracts.
- The overall growth of wireless voice usage in terms of new customers has slowed, and growth in revenue is even slower. The 1997 U.S. growth in subscribers slowed to 26%, while revenue growth was down to 16%. While the absolute growth in subscribers was still higher than that in 1996, revenue growth actually declined for the first time in recent years.
- No new services or features have significantly increased carrier revenues over basic air time and monthly fees since the new cellular generation of mobile phone service began in 1983.
- A carrier with 1 million subscribers, a churn rate of 2.5%, and a growth rate of 26% requires only 260,000 customers to grow at that rate, but 300,000 additional customers are needed just to replace churn. Many mature carriers are just reaching the point where the activations needed to replace churning customers exceeds their net gain in activations (The Steuernagel syndrome).
- Carriers are consolidating to expand their footprint and lower their costs to stay competitive.

Therefore, wireless voice carriers, especially mature cellular carriers, must look at what will sustain their growth as a business beyond the next three to five years.

Several applications have emerged as potentially important wireless products of the future. Many carriers have been bundling services in order to increase revenue per subscriber and customer loyalty. Obtaining multiple services from a known supplier is an important benefit for many customers, but a consolidated bill for multiple services is not always a positive—the bottom line becomes too big. Also, many providers who bundle services believe that the customer who needs one service will take a bundle at the same time. Unfortunately, customers are usually in a decision mode for only one service at a time.

It does happen to be a good time to add lines of business in telecommunications. There is great interest in Internet providers, alternative voice carriers such as Internet telephone and wireless local loop, local area networks wireless for data, the mobility offered by wireless voice for the office, and the ever-increasing need for bandwidth, whether wired or wireless. Many of these services are required by the same market segments that are currently using cellular and PCS. But offering such services might better be accomplished through reselling or acquisition, rather than building a network, in order to test the concept first.

It is extraordinary that most wireless carriers have not expanded their product offering to leverage their customer base with additional telecommunications offerings. Long distance, in connection with wireless service, is probably the only service in which they have been uniformly interested. The interest in bundling features is so great that the potential for offering additional unrelated telecommunications services to their customer base is ignored. These services will pose challenges to the sales and marketing organization of carriers, as well as technical challenges.

Wireless Office. Wireless carriers have the expertise and the leadership in the technology to make the wireless office a success with the help of vendors. The ability to allow transparent use of voice telephony in the office while mobile with the same handset makes the use of standard cellular phones for this purpose more attractive that proprietary technologies and unlicensed frequencies. Wireless communications can also be a backup for service continuity in case of a wireline service outage. Wireless local area networks in the office are an important addition to the product line. These opportunities also position the carrier for lucrative control of the account for public wireless voice and data service and increase the loyalty of existing corporate accounts on regular wireless service.

Wireless Local Loop. Wireless carriers have the opportunity to command this market, regardless of the availability of proprietary wireless local loop (WLL) technologies, which can service customers more inexpensively. The low cost of the current handsets will provide lower total cost per customer for some time. The success of the service will depend more on marketing than cost to serve. Local telephone companies need to be convinced to use this technology as a reseller, while wireless carriers can compete for the business themselves as regulatory authorities allow. Use of the phone outside the local home zone (within $\frac{1}{4}$ mile of the phone) can have premium air-time rates, whereas the phone works as the most versatile and longest-range cordless phone possible in the home zone at current wireline flat rates.

Existing wireless carriers are well positioned to offer local telephone service. They have many imagined disadvantages that prevent them from effectively competing with wireline local exchange providers:

1. *The price of wireless communications is too high.* Actually, competition is forcing down the price of wireless service to be competitive with wireline service. Besides, with the advantage wireless offers over wireline service, it can easily be competitive at a premium price. With subsidies, wireless phones are cheaper than wireline phones. Wireline service installation costs more than wireless setup and needs to be reinstalled every time a customer moves within the provider's service area.

2. *Wireless does not have the quality of wireline service.* Wireline service should not be the standard of quality in a competitive environment. If wireless service is not as dependable as wireline, it is dependable enough and has different advantages. For example, if wireline dial tone availability was measured to include moving vehicles, stranded motorists, mountain hikers, beach blankets, pedestrians, and boats, it would lose to wireless service.

3. *It costs too much to provide service.* It does not cost too much to provide wireless local loop service if you consider the marginal costs of an existing provider's excess capacity instead of the engineer's cost per customer served in new construction. If engineers were truly objective about the cost to serve a customer, wireline carriers would be buying service from wireless companies to serve customers more than a few miles from a central office.

4. *Wireless has the wrong sales channels.* This may be wireless communications's biggest problem. Surprisingly, basic wireline telephony customers are used to a provider with no physical presence and poor customer service. The wireless carrier needs to adapt to providing the kind of channel the basic service customer wants in the context of good service and cost—and may already have this capability in place. Strategies such as offering wireless service as the *second* line in a family residence or business will increase the

overall utility of such telecommunications service while lowering the wireless carrier's sales costs.

With its few problems, wireless communications has some incredible advantages over wireline telephone service:

1. Wireless service is not just mobile, it is available faster at new locations and in places in which it is not feasible nor cost effective for wireline service.
2. Standard wireline telephones are dumb, of poor quality, and featureless. Standard wireless phones have battery backup, display, memory, redial, are of higher quality, and have a host of additional features at a lower price.
3. Wireless phones have a larger toll-free calling area.
4. With wireless communications, more choice of pricing plans and structures, including prepaid and free long-distance calls and other elements not available in wireline service are possible.
5. More retail and service locations are available for wireless service.
6. Superior and competitive customer service, in person or by phone, is provided with wireless service.
7. Wireless communication is available immediately.

Dispatch. Wireless carriers have the capability to offer flat-rate service among users in the same calling group with distinct advantages over traditional dispatch services such as duplex mode and privacy. Some are already doing so. This service can also be an additional conduit to the wireless office.

Internet. Wireless carriers are a natural way to offer Internet services as resellers through other Internet providers using their current customer base. The carriers have a value-added role in making wireless access to the Internet transparent when clients are out of the office. The Internet connection is even more powerful as it is the entry point for emerging applications and telecommunications services of the future such as Internet voice, high-speed data, and video conferencing.

Carrier Behavior

Carriers must also change their attitude in order to become better wireless industry members and to enlarge the market for their services, by completing unfinished tasks. In addition to marketing additional services to their customers and others, they are blind to opportunities to expand their business because of their myopic view of the industry.

Resale. Wireless carriers must be a reseller in cases in which they do not provide service themselves. As carriers they should recognize the importance of the reseller channel for their own business. Most carriers are much too concerned with control of channels and product applications to be an effective value-added reseller or to support one. Becoming a reseller allows them to test their ability to market additional services before they try to build their own networks. It also permits them to offer services in areas in which they are not a carrier, so that their customers never need to be "roamers." Supporting resale provides revenue that has no sales cost and goes almost directly to the bottom line. It permits other providers to bundle wireless services into their own offerings. It makes local wireless carriers into national wireless providers.

Dealers and Retailers. Carriers have been fighting the need for third-party distribution since the beginning of cellular service. Carriers fight the cost of sales commissions even when the true cost of their own sales is higher. They introduce their own stores in the name of quality, when in fact it is an issue of control and phantom cost savings. The real issue is that consumers look for value where they expect to find it—at local dealers and consumer electronic stores, not just at carrier boutiques. Carrier stores may add value, but they cannot replace other channels.

Local Roaming. Carriers have shunned customers from other carriers in their own markets, believing they are keeping out competition. This may have been true in 1985, but today carriers need to permit reciprocal local roaming to increase revenue, reduce peak loading, and most of all avoid customer frustration with coverage holes and temporary call blockages. PCS carriers can especially benefit by cooperating with another carrier on the same standard to increase apparent coverage and reduce deployment costs.

Nationwide Pricing. AT&T's One Rate and its copycat offerings from other carriers are the first evidence of a real effort to integrate nationwide pricing and long-distance service. However, there is much more to be done. Carriers are still charging expensive long-distance rates for their regular subscribers to call between their markets, even adjacent ones. Wireless carriers have the opportunity to change the nature of long-distance telephony to be completely transparent, whereas their landline competitors subsidize monthly telephone access charges with stiff local and long-distance tolls. Not only should roaming rates be the same as for home users, but rate levels and structures should be more uniform from one market to another within the same carrier.

INDEX